EXTENDING OURSELVES

Extending Ourselves

Computational Science, Empiricism, and Scientific Method

PAUL HUMPHREYS

OXFORD
UNIVERSITY PRESS

2004

OXFORD

UNIVERSITY PRESS

Oxford   New York
Auckland   Bangkok   Buenos Aires   Cape Town   Chennai
Dar es Salaam   Delhi   Hong Kong   Istanbul   Karachi   Kolkata
Kuala Lumpur   Madrid   Melbourne   Mexico City   Mumbai   Nairobi
São Paulo   Shanghai   Taipei   Tokyo   Toronto

Copyright © 2004 by Oxford University Press, Inc.

Published by Oxford University Press, Inc.
198 Madison Avenue, New York, New York 10016

www.oup.com

Oxford is a registered trademark of Oxford University Press

Library of Congress Cataloging-in-Publication Data
Humphreys, Paul.
Extending ourselves : computational science, empiricism,
and scientific method / Paul Humphreys.
p. cm.
Includes bibliographical references and index.
ISBN 0-19-531329-1
1. Science—Computer simulation. 2. Science—Philosophy.   I. Title.
Q183.9.H82   2004
501'.13—dc21   2003054939

1 3 5 7 9 8 6 4 2

Printed in the United States of America
on acid-free paper

FOR DIANE

Felices ter et amplius
Quos irrupta tenet copula nec malis
Divulsus querimoniis
Suprema citius solvet amor die.
—Horace

Preface

This book originated in early 1989 during a conversation over the garden hedge with my neighbor, Jack Mitchell, who complained that computer simulations were forcing honest experimentalists out of business in solid state physics, an area to which he had made many significant contributions. Coincidentally, I had begun to notice here and there that parts of contemporary science were driven by computational methods and I had occasionally tried, without much success, to fit these new methods into the familiar categories of the philosophy of science. The coincidence revived my interest in the area and led, after some initial papers and a period of time prolonged by various practical distractions, to this book. The influence of computational methods on science today is far more striking than when that conversation took place, and many parts of science have, as a result, entered a distinctively different phase of their development. In this and in other areas, human abilities are no longer the ultimate criteria of epistemic success. Science, in extending its reach far beyond the limits of human abilities, has rendered anthropocentrism anachronistic.

Discussions in those early days with Fritz Rohrlich were of the utmost value in writing the first papers, as were, a little later, enjoyable and constructive exchanges with Stephan Hartmann. Generous contact with David Freedman's astringent intellect led over the years to a more realistic perspective on the limits of modeling. Conversations and correspondence with others resulted in significant improvements in the manuscript. I am especially grateful to Peter Achinstein, Jody Azzouni, Bob Batterman, Martin Carrier, Hasok Chang, Alexander Pruss, Robert Rynasiewicz, Tiha von Ghyczy, Mark Whittle, Bill Wimsatt, various academic audiences, and the anonymous referees from Oxford University Press. NSF grants

DIR-89-11393 and SES-96-18078 provided welcome support, as did two Sesquicentennial Awards from the University of Virginia. The book is dedicated to my wife, with gratitude for her help and forbearance.

The book contains in various places ideas that were originally presented in journal articles. All have been heavily reworked and are now, I trust, less tentative and better articulated than in their first appearances. I am, nevertheless, acutely aware of how incomplete is the treatment of many of the subjects discussed here. I wish I could describe the book in the delightfully arch words with which Descartes concluded his work *La Géométrie* in 1637: "But it is not my purpose to write a large book. I am trying rather to include much in a few words, as will perhaps be inferred from what I have done.... I hope that posterity will judge me kindly, not only as to the things which I have explained, but also as to those which I have intentionally omitted so as to leave to others the pleasure of discovery." But in the present case, one would need to replace "intentionally" with "unintentionally."

Contents

1 Epistemic Enhancers 3
   1.1 Extrapolation, Conversion, and Augmentation 3
   1.2 Science Neither by the People nor for the People 6

2 Scientific Empiricism 9
   2.1 Empiricism and Reliable Detection 9
   2.2 The Dilution Argument 12
   2.3 Accuracy, Precision, and Resolution 16
   2.4 The Overlap Argument 17
   2.5 Property Cluster Realism 22
   2.6 Bridgman Revisited 25
   2.7 Instruments as Property Detectors 28
   2.8 Know Thy Instruments 30
   2.9 Properties 41
   2.10 Epistemic Security 45

3 Computational Science 49
   3.1 The Rise of Computational Science 49
   3.2 Two Principles 55
   3.3 Units of Analysis 57
   3.4 Computational Templates 60
   3.5 "The Same Equations Have the Same Solutions":
      Reorganizing the Sciences 67
   3.6 Template Construction 72
   3.7 Correction Sets, Interpretation, and Justification 76
   3.8 Selective Realism 82
   3.9 Further Consequences 85
   3.10 Templates Are Not Always Built on Laws or Theories 88
   3.11 The Role of Subject-Specific Knowledge in the
      Construction and Evaluation of Templates 91

3.12 Syntax Matters 95
3.13 A Brief Comparison with Kuhn's Views 100
3.14 Computational Models 102

4   Computer Simulations 105
    4.1 A Definition 105
    4.2 Some Advantages of Computer Simulations 114
    4.3 The Microscope Metaphor and Technological Advances 116
    4.4 Observable, Detectable, and Tractable 121
    4.5 Other Kinds of Simulation 125
    4.6 Agent-Based Modeling 129
    4.7 Deficiencies of Simulations 133

5   Further Issues 137
    5.1 Computer Simulation as Neo-Pythagoreanism 137
    5.2 Abstraction and Idealization 141
    5.3 Epistemic Opacity 147
    5.4 Logical Form and Computational Form 151
    5.5 In Practice, Not in Principle 153
    5.6 Conclusion 156

References 157

Index 169

EXTENDING OURSELVES

1

Epistemic Enhancers

1.1 Extrapolation, Conversion, and Augmentation

Let us suppose that we were all equipped with sensing devices which could detect every kind of object and property that exists in the material world, small, large, medium-sized. Suppose also that we had mathematical skills even greater than those with which Laplace endowed his famous superhuman agent, so that not only were calculations easy and error-free but the construction of exceedingly complex mathematical models came as naturally to us as do simple arithmetical representations. We would, in short, be epistemological gods.[1] And not just epistemological deities. It has been said that it is in the realm of perfect knowledge that metaphysical truths most clearly emerge. If so, then the constituents of the world would be directly accessible to us. In these circumstances would we have any use for science? Precious little, I should think.[2] Once science has perfected this extension, it will have put itself out of business. The end of science achieved would be the end of science.

One of the things to which this fantasy draws our attention is the fact that one of science's most important epistemological and metaphysical

---

[1] Not God, of course. You could have all this and be morally imperfect. The thought exercise is not meant to be taken literally, for such a scenario might well be physically impossible in the sense that there are insuperable physical limitations on certain types of calculations which require unbounded amounts of memory.

[2] The little might well be precious, for it would probably include the need for theories of space-time or of less mundane spaces, which would not, one imagines, be detectable even by such gods. Scientific theories might also unify what we know. But in the scenario described, we should have direct access to all the fundamental forces of nature and to all the causes of natural and social phenomena, and so such unification might well be unnecessary. Whether there would be a need for the social sciences I leave as an exercise for the reader.

achievements has been its success in enlarging the range of our natural human abilities. Sometimes this enhancement takes place by extending an existing modality, such as vision, along a given dimension. We can call this *extrapolation*. This kind of enhancement is most familiar in the case of instruments such as optical telescopes and microscopes, where the very faraway and the very small are brought within the range of visual detection. In other cases of extrapolation we do this by expanding the domain of our existing abilities, as when the limited range of electromagnetic radiation that we can visually detect is expanded to include most of the electromagnetic spectrum, using infrared and millimeter wavelength telescopes. An alternative method of enhancement is *conversion*. This occurs when phenomena that are accessible to one sensory modality (perhaps through enhancement in one of the above two ways) are converted into a form accessible to another, such as with sonar devices that have visual displays. Perhaps most important, it gives us access to features of the world that we are not naturally equipped to detect in their original form, such as alpha particles, positrons, and spin.[3] We can call this last kind of enhancement *augmentation*.

Extrapolation involves cases where our senses allow us access to certain values of a quantitative property and the instrument extends that access to other regions of the same property. Examples of extrapolation are the use of instruments that detect sound waves outside the usual realm of 20 Hz to 20,000 Hz that is accessible to human ears, and ultraviolet radiation detectors that extrapolate along the continuum of electromagnetic radiation from the region to which we have native access. A special case of extrapolation occurs when instruments allow us to have access to lower-intensity instances of the same regions of the same properties that are accessible to the unaided senses. For example, optical telescopes operate on visible wavelengths of the electromagnetic spectrum, just as our eyes do. The telescopes are simply capable of gathering more light than are our unaided eyes. We can include such cases within extrapolation without prejudicing our later conclusions, although the conceptual difference between the two is important. Conversion has already been explained, and is of limited use except in conjunction with one of the other two methods.

---

[3]For an excellent collection of images illustrating such extensions, see Darius 1984. Arguments for realism based on the idea of humans equipped with different sensory abilities can be found in Maxwell 1962. Related arguments can be found in Churchland 1985, with responses in van Fraassen 1985.

Augmentation involves cases, such as the detection of magnetism, where no values of the property naturally affect our sensory organs, the property being of a qualitatively different kind from those which do.

Such enhancement is not limited to perceptual abilities. Our natural mathematical talents have been supplemented by computational devices that enable us, in certain areas, to move beyond what our psychological capacities can achieve. At present, this mathematical supplementation is almost exclusively limited to extrapolation and conversion, for it primarily covers analogues of only two of the three sensory enhancements with which I began. It can greatly increase the speed with which we can perform certain mathematical operations, thus altering the time scale in ways analogous to the way that optical telescopes alter spatial scales. It can expand the domain of problem complexity, allowing us to move from studying a restricted range of simple topological structures to structures of far greater complexity or of different types. It can convert numerical results into graphical form, as is often done in statistical analysis, allowing easier access to massive data sets by shifting from one representational mode to another. Observational and computational enhancements complement one another when computationally assisted instruments are used, with physical devices and mathematics working together to give us increased access to the natural world. An example of this is computerized axial tomography (CAT scans), within which physical detectors pick up differential rates of radiation and the output of those detectors is transformed by mathematical algorithms into two- or three-dimensional images of the object under investigation.

What rarely happens is an analogue of the third type of empirical extension, augmentation, giving us instrumental access to new forms of mathematics that are beyond the reach of traditional mathematical understanding. Here the question naturally arises: Could instrumentally enhanced mathematics allow an investigation of new areas of mathematics to which our only mode of access is the instrument itself? It is a question of the first magnitude whether this would even be possible, but it is not one that I shall discuss in detail here. I shall merely note that we already have one such extension in formal proofs.[4] Even in their

---

[4]'Formal' is intended here to be roughly synonymous with 'explicit and rigorous,' where the proof is carried out with the aid of some concrete medium. The term is not intended to be restricted to purely syntactical proofs, and proofs construed as abstract objects do not have the desired instrumental features.

elementary pen-and-paper form, these are instruments that extend our native inferential abilities in ways which, at least, supplement our memory capacity. Their use as augmenters occurs when mathematical intuitions and formal proofs are at odds with one another, because in those cases proofs usually override the intuitions. Unlike the senses, intuitions can often be adjusted in the light of instrumental outputs, in this case the structure and conclusions of the proofs. Of course, in the other direction, intuitions can provide insights that cannot be immediately proved, but that situation is parallel to the case in which observations provide a stimulus for novel discoveries in the nonobservational realm.

Our task as philosophers is to identify the ways in which these new methods are changing the way in which science is done.

1.2 Science Neither by the People, nor for the People

Automatic telescopes represent a novel concept leading to a radically new way of planning and conducting observations. This is best illustrated in photoelectric photometry where the human factor is one cause of errors. Man, with his slow reaction time and high proneness to fatigue, cannot compete with a computer and with ultrafast equipment.[5]

These considerations seem to suggest that one of the principal achievements of science has been to transcend the limitations of humans' natural epistemic abilities. But could human participation be avoided entirely? Consider a scenario, not too far into the future, when humans have become extinct. A rocket that was constructed before mankind perished is launched, and its trajectory, which is automatically computed by an onboard device, takes it to a distant planet. The rocket contains an automated breeding module, and when it arrives at the planet, it samples the atmosphere, the temperature, the available nutrients, the terrain, possible predators, and so on, all of which were unknown and unpredicted by humans. It then engineers the genetic code for a new cognitively aware species that will survive and flourish in the alien environment. The species colonizes the planet and adapts it to its own purposes. This all happens without a single human knowing anything specific about how that part of the universe has been controlled by science. Yet this is still science,

[5]Sterken and Manfroid 1992, p. 48.

indeed scientific knowledge, in operation. This imagined scenario is merely an extension of what presently occurs with the robotic synthesis of molecules, with automated DNA sequencing, with the automatic detection of particles in accelerators, and with the use of automated data collection and reduction in some areas of astronomy.[6] The moral of the story is that an activity does not have to be done either by us or for us in order to be considered scientific.

There is a different aspect of automation, which we may call *the quantity of data issue*. A standard part of scientific method requires us to collect data and to compare those data with the predictions of our theories or models. In many areas of science it is completely infeasible to do this 'by hand'. The calculations to perform the shotgun approach to mapping the human genome involved 500 million trillion base-to-base comparisons requiring over 20,000 CPU hours on the world's second largest networked supercomputer, a task that is claimed by the company involved to be the largest computational biology calculation yet made. The database used contained eighty terabytes of DNA analysis, equivalent to a stack of manuscript sheets 480 miles high, and more than five times the amount of syntactic information contained in the Library of Congress. Another area, turbulent flow research, which is required for applications in areas as diverse as the flow of air over an aircraft wing and the detailed representation of a nuclear weapons explosion, is one of the most computationally intensive areas of science. But data production in this area is equally impressive. Fluid flow data can now be recorded every 10^{-3} seconds and, to quote, "The technology of measurement [in turbulent flow research] is probably undergoing a faster explosion and evolution than the technology of computation."[7] In astronomy, vast quantities of data are accumulated by automated instruments. The Diffuse Infrared Background Experiment on the NASA Cosmic Background Explorer satellite, for example, produces 160 million measurements per twenty-six week period.[8]

For such data sets, the image of a human sitting in front of an instrument and conscientiously recording observations in propositional form before comparing them with theoretical predictions is completely unrealistic.

---

[6] See, e.g., Moor et al. 1993 for background information on molecular synthesis; Henry and Eaton 1995 for details of automated astronomy; and Adelman et al. 1992 for some cautions on the current limitations of such methods in astronomy.

[7] See Cipra 1995.

[8] E.L. Wright 1992, p. 231.

Technological enhancements of our native cognitive abilities are required to process this information, and have become a routine part of scientific life. In all of the cases mentioned thus far, the common feature has been a significant shift of emphasis in the scientific enterprise away from humans because of enhancements without which modern science would be impossible. For convenience, I shall separate these enhancements into the computational and the noncomputational, beginning with the latter. This division of convenience should not conceal the fact that in extending ourselves, scientific epistemology is no longer human epistemology.

2

Scientific Empiricism

2.1 Empiricism and Reliable Detection

Epistemic enhancers promise to radically widen the domains of what counts as scientifically observable and mathematically tractable. Beginning with the former, the division between what is observable and what is unobservable has traditionally served as the dividing line between what is acceptable as evidence to an empiricist and what is not. One reason, although certainly not the only reason, why empiricists have wanted to maintain a sharp division between the observable and the unobservable is their desire for epistemic security. Direct perceptual knowledge of the observable might be defeasible, they reason, but it does not have the shaky status of our abandoned beliefs in phlogiston, the superego, noxious effluvia, and all the rest of those entities which turned out to be unobservable because they did not exist. This concern for safety is entirely legitimate, and our first step in developing a scientific empiricism should be to take the security issue as primary. Drawing a distinction between what is observable and what is unobservable is merely one attempt to resolve this epistemic security problem. That is, it is not the intrinsic status of something as 'observable' or 'unobservable' that should concern us, but how good the detectors we use are at producing secure knowledge.

In that respect, the traditional human sensory modalities are neither particularly reliable nor all that versatile. Here are two test cases.

1. You were an eyewitness to a crime that you observed in broad daylight, under 'ideal conditions'. At that time you were certain, based on your visual identification, that the individual committing the crime was your best friend, whom you have known for years. The assailant fled, bleeding profusely from a wound. Immediately

afterward, in your job as an FBI lab technician, you personally collect blood samples at the scene and run a DNA analysis on the samples. The analysis excludes your friend as the source of the blood with a probability of error of less than 1 in 10^9. Which evidence would you accept?

2. The label has disappeared from a container in your garden shed. You smell the liquid inside and swear that it is Whack-A-Weed, an herbicide you have been using for two decades. Just to be sure, you run the liquid through a molecular analyzer at your lab. It identifies the liquid as Old Yeller, a whiskey you kept in the shed for cold days. Which evidence would you accept?

The answers are, I hope, obvious. No methods of detection are perfectly reliable, but barring strange chains of events, instruments trump human senses here, there, and almost everywhere. This superiority of instruments over human senses is widely recognized. Photo finishes in horse races, electronic timing in Olympic swimming and sprint events, touch sensors in fencing, automatic pilots, light meters for camera exposures, hygrometers to test soil dampness, and so many other examples are all reflections of this superiority. Readers of a certain age may remember the need for a 'personal equation' as a corrective for human observers in scientific experiments such as scintillation counts before the era of automatic detectors. Many of the optical illusions that fool the human visual apparatus do not affect computerized image analyzers.[1] Given this widespread preference for instruments, why should empiricists have reservations about their use? In order to understand their hesitation, and to see why these reservations are misguided, we need to consider what constitutes empiricism and the reasons why some philosophers have been attracted to it.

A variety of approaches have gone under the name of empiricism, and the motivation for defending it differs between adherents. So perhaps it is unwise to try to define the term.[2] Still, there does exist a core content to traditional empiricism. Here is one representative definition:

> In all its forms, empiricism stresses the fundamental role of experience.... It is difficult to give an illuminating analysis of

---

[1]See Russ 1990, p.12.

[2]In the course of writing this section, I came across a book promisingly titled *The Encyclopedia of Empiricism*. It had, disappointingly, no entry on empiricism. It is, I belatedly realized, a common feature of reference works devoted to X that they never think to tell the reader what X is.

'experience' but let us say that it includes any mode of consciousness in which something seems to be presented to the subject....
Experience, so understood, has a variety of modes ... but empiricists
usually concentrate on sense experience, the modes of consciousness that result from the stimulation of the five senses.[3]

The twentieth-century empiricists who drew a boundary between
observable entities and unobservable entities (or between the classes of
terms that designate them) intended that boundary to be permanent. If
rabbits are observable, they always have been observable and always will
be.[4] And because the criterion of observability was tied to ideal observers
in ideal conditions, there was in this tradition no difference between
observability in principle and observability in practice. Such empiricists
tended to stay very close to our unaided abilities, insisting that evidence
from the five traditional senses must be the ultimate arbiter in evaluating
claims to knowledge. This sort of traditional, anthropocentric empiricism
is not well suited to the needs of science. Science makes claims, claims
that are often highly confirmed—dare I say even true?—about subject
matter that goes well beyond the reach of our natural senses. Yet even
versions of empiricism that are designed to be sympathetic to the special
needs of science remain grounded in the human senses. For example, van
Fraassen characterizes his constructive empiricism in these terms:

> Science presents a picture of the world which is much richer in
> content than what the unaided eye discerns. But science itself
> teaches us also that it is richer than the unaided eye *can*
> discern.... The distinction, being in part a function of the limits
> science discloses on human observation, is an anthropocentric one.[5]

This fixed, sharp division between the observable and the unobservable fails to capture important differences in the nature of the subject
matter and its relationship to us, for one of the most striking features of
what counts as observable in science is that the dividing line between the

---

[3]Alston 1998, p. 298. Alston's definition allows pain, for example, as acceptable to some
empiricists, although we shall not pursue that issue here.

[4]Individual entities might pass from being observed to being unobserved—a comet in a
hyperbolic orbit could be observed when close to Earth but could not be observed directly
when it was outside the solar system—but since there are conditions under which such
entities can be observed, it is legitimate to consider them to be permanently within the realm
of the observable.

[5]Van Fraassen 1980, p. 59.

observable and the unobservable is not fixed but is moving, with the trajectory being determined in part by technological advances. Constructive empiricism is a sophisticated and enlightening epistemological position.[6] Nevertheless, these residual influences of anthropocentrism are unnecessarily constricting. Suppose for the moment, however, that we accept that there is a fixed line between what is observable and what is unobservable. Even then, traditional empiricists ought to allow the existence of certain things that are usually classified as unobservable, as we shall now see.

2.2 The Dilution Argument

We can capture the essence of traditional accounts of empiricism in this idea: An entity is observable just in case there exists a nomologically possible situation within which the relationship of that entity to an ideal human observer is such that the entity is directly observed by at least one unaided human sense possessed by that observer. Technological aids are permissible in creating that nomologically possible situation as long as the aids are not involved in the observation itself and do not alter the entity in creating that situation. Using a mechanical excavator to expose an underground water pipe allows us to observe the pipe, thus showing it to be observable, and so does riding a rocket to the Moon to inspect what lies just beneath the surface. Turning on the electric light in a darkened room to check the damage the dog did counts just as much as waiting for the Sun to rise before making an inspection, as does using a microwave oven to defrost the stew so that we can taste it. Such cases are surely acceptable to an empiricist, and they reveal something important about the concept of an observable.

What the cases cited above have in common is that an unobserved object or property has been moved across the boundary between the unobserved and the observed (not across the observable/unobservable boundary) so that it is now available for direct inspection. This boundary, it is important to note, is, for traditional empiricists, fixed. That is, the entity, when unobserved, possesses the dispositional property of being observable because it is nomologically possible to create a situation within

---

[6]Van Fraassen's most recent book on empiricism (2002) was published after the manuscript of this book had been completed. I regret that I have been unable to address some of the issues he raises there.

which it is observed, 'observed' here being taken in the traditional, minimalist, sense. But if the entity has that dispositional property when it is not being observed and it is acceptable to empiricists for it to retain the label 'observable' when it is moved across the unobserved/observed boundary from the realm of the unobserved to the realm of the observed, some things ought to retain that label when they are moved across the boundary in the opposite direction, from the realm of the observed to the realm of the unobserved. So, we bury the pipe and believe that it continues to exist; we leave the Moon to return to Earth and hold that the subsurface lunar composition remains the same; we turn off the light and regret that the dog's damage remains to be repaired; we refreeze the leftover stew, expecting that tomorrow we can taste it again, with the additional expectation that it will taste different.

It is this movement back and forth across the boundary between the observed and the unobserved that is important, because if we have good grounds for believing in the continued existence of certain kinds of entities even when they are in the realm of the unobserved, then the use of instruments other than human observers can justify claims about the existence of certain kinds of entities that never enter the realm of the humanly observed. This justification can be made using *The Dilution Argument.* Take 1 gram of finely ground iron dust, the totality of which when heaped or when scattered on a white surface can be observed. Now mix the dust thoroughly into a bucket of black sand. The iron in that situation cannot be observed by any of the unaided human senses, including the visual, but we can verify its continued existence by weighing the bucket before and after the addition of the iron, using a sensitive balance. If the use of a conservation of mass principle to underpin the continued existence claim is questioned, the existence claim can be buttressed by recovering the iron from the sand by using a magnet and demonstrating that within the limits of experimental error, the original quantity of iron has been recovered. Here, the mere movement of a substance back and forth across the visually observed/visually unobserved boundary does not in itself undermine our justification for its continued existence.

Thus far, we are still within the realm of extrapolation of our native abilities, because we can (crudely) compare the values ourselves for objects of moderate weight and the balance is simply an extension of that native ability. But we can move into the realm of augmenting our senses by magnetizing the iron before mixing it. Then, a magnetometer run over the contents of the bucket before and after the addition of the iron will show a differential effect, the only plausible explanation for which is the presence of the iron, the existence of which within the sand has already been established on the basis of direct inspection within the realm of the visually observed followed by dilution. In related situations—and all the interesting complexities and details are hidden in the 'related'—we can now, with the use of augmentations of our native senses, reliably detect the presence of iron in sand.[7] Then, in other cases where the iron has never been observed by our own sensory apparatus, we can justify claims for its existence on the basis solely of instrumentation. It is this ability to make existence claims entirely on the basis of instrumental evidence rather than on the basis of a hypothetical, unrealized potential for unaided human observation that should incline us to abandon the restricted perspective of traditional empiricism, because it is not the unmanifested disposition of being humanly observable that plays a role in our judgment that this instance of iron exists, but the use of the instrument.

Unfortunately, the Dilution Argument can only be a subject-matter-specific argument for the existence of unobserved entities and not a general argument. For example, it cannot be used even in the apparently simple case of dissolving table salt in water. And in fact, that case reveals an important limitation on empiricism. Take an observable quantity, say 1 gram, of a water-soluble substance such as salt. Now dissolve the salt in 1 liter of water. The salt has passed into the realm of visually unobserved quantities, but it is still, apparently, accessible to the human taste modality. Although the salt is visually unobserved, its continued presence can apparently be verified in a manner acceptable to an empiricist. We can also return the salt to the realm of the visually observed by evaporating the water. Yet in this case, the empiricist inference would be wrong. For chemistry tells us that when salt is dissolved in water, it splits into sodium and chlorine ions and no longer exists qua salt, even though we cannot distinguish by taste between salt and salt solution. Indeed, as Hasok Chang pointed out to me, the human tongue cannot taste salt unless it is dissolved in saliva. And so the empiricist identification of salt as salt via the use of our unaided taste modality is itself wrong—taste buds identify a solution of sodium and chlorine ions and not salt.

---

[7]This argument assumes that the instrument is uniquely sensitive (i.e., there is only one kind of causal trigger for the instrument). When that assumption is not satisfied, the argument becomes more complex.

A more whimsical example is this: I have a mug upon the outside of which is an image of the Cheshire Cat. All but the cat's smile is rendered in heat-sensitive paint of the same color as the mug, so that when the mug is filled with hot liquid, the cat fades away, leaving visible only the smile. As the liquid cools, the cat returns. Is it correct to say that the cat image is still there even when invisible and that it has been moved back and forth across the observed/unobserved boundary by changes in temperature? That is a hard question, and the reason it is hard reflects a difference between this case and the case of the buried pipe. When entities pass back and forth across the observed/unobserved line—whether it is by dilution, by spatial distancing, by temperature changes, or by other means—we are implicitly using an identity assumption to infer their continued existence, an assumption that is sometimes satisfied, sometimes not. In the case of the buried pipe, the identity criteria are unproblematical; in the cases of the invisible Cheshire Cat and the salt, they are not. What is important to know is whether the processes that take an entity from the realm of the observed to the realm of the unobserved, or vice versa, are such that the identity of that entity is preserved. The usually straightforward nature of the process in cases where one merely moves closer to an object in order to observe it is unrepresentative of the difficulties involved in more problematical cases. The need for this kind of specific knowledge about identity means that there can be no general argument for the existence of unobservables on the basis of their persistence across an observed/unobserved boundary, and to search for one is likely to be a fruitless endeavor.

The Dilution Argument shows that this identity assumption is hidden in some empiricist arguments for the existence of observables. Take, for example, van Fraassen's attempt to maintain a balance between the demands of science and the anthropocentric origins of empiricism. For him, the moons of Jupiter count as observable because we, as presently constituted humans, could see them directly from close up, were we in a position to do so.[8] This use of technology is presumably legitimate because it would leave both our visual sense and the target objects unaltered. But the identity assumption is employed here as well. It is tacitly assumed that what we would, counterfactually, see directly are the same objects we see through a telescope. If we consider how we test such identity claims, we often employ a second kind of argument that I shall call the *Overlap Argument*. Hypothetically, one could travel to Jupiter while keeping one eye trained on its moons through a telescope and observing directly, with the other eye, those moons grow ever larger. The overlap between these simultaneous observations should convince us that indeed the same objects are given to us by the telescope and our unaided vision. All of this would take place within the realm of what van Fraassen considers as observable and so should be considered unobjectionable to a traditional or constructive empiricist. However, the general form of such overlap arguments will allow us to make the case that the boundary of the observable is not fixed but is continually being pushed back as our instrumental access to the world improves. Before we turn to that argument, we need to discuss some relevant concepts.[9]

2.3 Accuracy, Precision, and Resolution

There are three aspects to detection that we may call accuracy, precision, and resolution. An instrument producing quantitative data is accurate on a quantity Q if its output closely approximates the true values of Q; an instrument has a high degree of precision on a given value of its output if the empirical variance of the output on that value is small. (One may subdivide this into individual precision and collective precision, with the former being concordance of data from a single observer and the latter being the reproducibility of data from different observers.) One instrument I_1 has superior resolution to another instrument I_2 with respect to some set of values S of the quantity Q and some metric on those values if, when used to measure Q, I_1 has at least the same resolving power on Q over all pairs of values in S as does I_2 and at least one pair of values in S with greater resolution.[10] That is, I_1 can distinguish all the elements of S that I_2 can distinguish and at least one further pair of elements in S.

These features of an instrument are largely independent. Accuracy entails precision in the sense that a sequence of accurate data drawn from the same source will be precise but the instrument used need not have high resolution. Nor does high precision entail accuracy of the individual data points—systematic bias in a precision instrument can wreck its accuracy. Conversely, an instrument with high resolution may be neither accurate nor precise. Optical Character Recognition (OCR) software used

---

[9]Criticisms of an earlier version of this argument by Martin Carrier and Alexander Pruss were instrumental in improving this section.

[10]This can be applied to qualitative properties by applying a discrete measurement metric to categories of the property.

with scanners almost always gives the correct number of characters in a string, thus having good resolution regarding where one character ends and another begins, but early versions of OCR software had very poor precision and accuracy. Running the same text through the software multiple times would result in a high variance in recognizing a given token of the letter 'u' , for example. In contrast, if resolution involves character types rather than character tokens, then good resolving power on types does entail high precision on tokens, although not high accuracy. Perhaps accuracy is the most important virtue for empiricists—they do not want to be mistaken about matters of fact in the realm of observables—but precision without accuracy could well be important to an anti-realist empiricist.[11] In general, empirical realists should want their instruments to have all three virtues. One accurate detector is then preferable to another if the first has resolution superior to that of the second.

We know that particular detectors have different domains of accuracy and resolution and that those domains are often quite limited in scope. As observers, we humans are subject to constraints that preclude us from having precise, accurate, and complete information about most systems and their states. So let us use the idea that human sensory capacities are merely particular kinds of instruments, sometimes better but often worse than other detectors. Human observers have fair accuracy within certain restricted domains; for example, if a flying ant and a Southern termite are compared in good light, the correct classification of each can repeatedly and accurately be made by many different observers with minimal training, thus making unaided vision an accurate detector of termites. Yet, as we have seen, despite the empiricists' elevation of human detectors as the ultimate standard, the domains of accuracy, precision, and resolution for the human senses are highly restricted, and even within those domains humans often do not provide the best results.

2.4 The Overlap Argument

Much of the debate about observables has taken place in terms of examples where we extrapolate our existing sensory inputs by using instruments such as microscopes, telescopes, and other artifacts. This is undeniably

---

[11]'Anti-realist empiricist' is not pleonastic. It is true that most empiricists tend to be anti-realists, but one does not want to trivialize the debates between empiricists and realists by ruling out by definition the possibility of reconciling the two.

interesting and important, but a characteristic feature of scientific instruments is their ability to put us into contact with properties of objects that are natively almost or completely inaccessible to us—the augmentation case. X-ray telescopes, functional PET imaging, electron microscopes, scanning tunneling microscopes, acoustical microscopes, auto-fluorescent bronchoscopes, and a host of other instruments detect features that lie far beyond the reach of our native senses. Indeed, one of the misleading conclusions that tends to be drawn from the discussions in the philosophical literature about microscopes and telescopes is that problems of observation are simply scaling problems. Extending ourselves involves gaining access to qualitatively new kinds of features as well as to more of the same kind. Even though cases of augmentation and conversion are more difficult for a scientific realist to defend than are cases of extrapolation, overlap arguments can be used to justify all three kinds of extension.[12]

Examples of overlap arguments are familiar to all of us. We trust what we see through a low-powered telescope because its domain of applicability overlaps what we can see with the naked eye, and within the visual domain the telescope reproduces accurately, precisely, and with good resolution familiar features of the world. I see my wife on the plain over yonder, and when I train the telescope on her, there she is again, simply larger and clearer. I can just see the fruit fly on the microscope slide— I put it under the low-powered microscope and here it is; I can see the head in much more detail through the greater resolution of the microscope.

How such arguments are employed is illustrated by a well-known philosophical argument for realism, first laid out explicitly by Grover Maxwell, that tacitly relies on just such an overlap argument. The point of his 'continuum of observation' argument is to convince the reader that any division between the 'observable' and the 'unobservable' is arbitrary, and, as presented, it fits naturally into the category of extrapolation:

> ...there is, in principle, a continuous series beginning with looking through a vacuum and containing these as members: looking through a windowpane, looking through glasses, looking through binoculars, looking through a low-power microscope, looking through a high-power microscope etc., in the order

[12]The augmentation case is more open than the others to what Collins has called the experimenter's regress (see H. M. Collins 1985, chaps. 4, 5) For responses to Collins's argument, see Franklin 1990. The extrapolation case is not subject to this problem because of our direct access to the property.

given.... But what ontological ice does a mere methodologically convenient observational-theoretical divide cut?... is what is seen through spectacles a "little bit less real" or does it "exist to a slightly less extent" than what is observed by unaided vision?[13]

Maxwell's argument was designed to persuade by immediate appeal, but the general form of the overlap argument that underlies it is as follows. We choose a source, some system possessing a property, such as shape, that is accessible both to the unaided senses and to the instrument. If the relevant property of the source is reproduced with sufficient accuracy, precision, and resolution by the instrument, this validates it within the region for which human senses serve as the standard. The key issue is then how the use of the instrument is justified outside the domain of application of the human senses. In this section I shall consider only cases for which a supplementary dilution argument that relies on the ability to move entities in and out of the initial range of observation can be brought to bear. So, my wife wanders farther away, becomes an indistinct object, and the telescope restores her image accurately and highly resolved. A speck on the ocean appears, I see with the telescope that it has the configuration of a barque, and eventually it cruises near enough for me to see it directly. This kind of overlap argument lay behind the quick acceptance in 1895–1896 of the earliest X-ray photographs. Skeletal structures that were revealed by the X rays could be moved from the realm of the unseen and unfelt into the domain of human vision and touch by anatomical dissection or by surgery.

These examples all involve extrapolation of our native senses. The iron dust example of the last section illustrates how an overlap argument can justify the use of certain instruments that augment our senses. Take the case where the iron dust has been magnetized. When the dust is visible, it provides the only explanation for the detection by the magnetometer of the associated magnetic field. Having established the accuracy of the magnetometer on that source and in the domain of the visible, the instrument's use is extended, by means of the dilution argument, to cases where the source is invisible within the black sand and also is inaccessible to all of our other native senses. The instrument in this case provides information about a property of a system that reflects a genuine augmentation of our senses. In other cases it will take us outside the realm of the humanly observable altogether.

---

[13]Maxwell 1962. Reading 'in the order given' literally, this passage is odd. Presumably Maxwell meant to indicate here a succession of 'distances' from unaided vision.

The fact that regions of overlap can occur between two instruments allows one of the most important features of these arguments to be brought into play. Once standards have been set for some new, extended domain of application, the overlap argument can be iterated by using that domain as the new standard to justify the use of other instruments that extend our reach even farther.

Thus far we have taken the traditional empiricist's view that ideal human observers serve as the initial calibration devices. Nevertheless, we have already seen that using humans as the fundamental reference point not only is highly restrictive but also constitutes bad practice in many cases. Humans do serve as the starting point for some phenomena, but they are often quickly abandoned and better detectors are employed for the origins of the overlap process. Crude tests for acidic solutions can be made by humans on dilute samples by taste. For base solutions, a soapy feeling, resulting from the dissolution of a layer of skin, allows a rough test. But these are limited and unreliable methods, and have been replaced in turn by litmus tests, phenolphthalein tests, reaction rate tests, conductivity tests, and other tests of far greater reliability. The calibration is now made on solutions of standard pH values, independently of human perceptual standards. Rough comparisons of temperature can be made by the human body, with a hand on the forehead of a child, but we are all familiar with how inaccurate, imprecise, and crude this can be. Most of us as children participated in the experiment where, with eyes closed, you first place your left hand in hot water and your right hand in cold water, then immediately place both hands in lukewarm water. The resulting conflict in judgments of the temperature of the lukewarm water is, sometimes literally, an eye-opener. Touching a metal surface and then a plastic surface on a cold morning results in radically different and incorrect judgments of temperature. So reliable temperature detection and measurement are delegated to thermometers of various kinds, each of which is calibrated on fixed points, either directly or indirectly, without appeal to validation by the human senses.[14]

The calibration of precision instruments frequently relies on an overlap argument, and although the procedures employed are far more delicate than those we have discussed, they are not different in principle except for the fact that explicit attention is paid to perfecting the calibration

---

[14]For a detailed account of the historical and philosophical difficulties that were involved in establishing thermometric fixed points, see Chang 2004.

standard, a procedure that is avoided in the usual appeal to ideal human observers.[15] In CCD detectors for astronomical photometry, for example, each pixel varies slightly in its sensitivity, and so a flat field calibration is made before observations are taken in order that corrections can be made on the ensuing images to compensate for variations in the detected brightness. In photographic plates, there is an S-shaped curve of sensitivity, with objects below a certain brightness not registering at all and objects above another level suffering from saturation of the image. So in the corner of each plate for astronomical use there is a reference set of images of standard brightnesses. Many satellite-based telescopic instruments are calibrated on the ground in laboratories where the nature of the radiation is known precisely. In cases where the calibration standard is not met, the evidence of the instrument is simply rejected. Buy a cheap telescope and—look—the porch light down the street appears double. Clearly an instrument not to be trusted when observing Sirius.

Often, the instrument will have resolution superior to that of the human senses within the region of overlap, but not always. It can have degraded reliability as long as that inferiority is taken into account when using the instrument outside the calibration range. For example, current side-scanning sonar detectors are worse than unaided human vision on all three counts of accuracy, precision, and resolving power, but in regions in which humans cannot operate, typically at extreme ocean depths, their use to detect underwater features such as shipwrecks is considered legitimate because the identification of gross properties such as the shape of a ship's hull can be made with the relatively poor quality they provide.

Only within the domain of overlap can the accuracy and resolution of the instrument be justified directly by comparison with the standard. The precision of the instrument, in contrast, can be determined across the entire range of its use. Curiously, we can sometimes validate instruments, such as low-powered magnifying glasses, by using less than ideal human calibrators. Those of us whose vision is not what it used to be can no longer resolve symbols in a familiar text, yet we remember that we once could. When the magnifier restores the degree of resolving power we

[15]Allan Franklin (1986, 1990, 1997) has stressed the importance of calibrating physical instruments to validate them as sources of reliable data. The importance of his idea extends to a much wider context than the sort of highly sophisticated research instruments in physics with which he is chiefly concerned. I am grateful to Mark Whittle of the University of Virginia Astronomy Department for conversations about the telescopic calibrations discussed in the rest of this paragraph.

know that we once had, we are willing to defer to the output of the simple instrument. This is a different argument than the basic overlap argument used to justify the use of magnifiers to those with 'perfect' vision.

The contrast between the cheap telescope and the side-scanning sonar is instructive. Both are inferior in certain ways to human vision, but whether their respective deficiencies are disabling depends upon the application. The inability to determine the correct number of point sources of light is fatal to most optical telescopes because images of stars are very often found close together. When sonar images are similarly blurred, it does not matter to the same degree for individuating the sources because shipwrecks are rarely found close to one another. In contrast, counting the number of cables on a ship presents serious difficulties.

What of cases where a dilution argument cannot be used? In such cases, which constitute the great majority of instrument applications, knowledge of the instrument and of the subject matter to which it is applied is necessary, a subject to which we turn in section 2.8. It is worth noting that because the dilution argument also requires subject-specific knowledge to be applied, the constraints are similar in each case but we require more knowledge of the instrument when dilution arguments are not available. Before we discuss the role played by knowledge of the instruments in overlap arguments, we must assess what, exactly, it is that instruments detect.

2.5 Property Cluster Realism

One obstacle to understanding this modality is that, while an NMR [nuclear magnetic resonance] image does depict physical properties of the subject imaged, these properties differ considerably from the properties imaged in other modalities. It is not a simple matter to say clearly what an NMR image is an image *of*. To complicate matters further, several factors affect image contrast. By varying the conditions under which NMR data are acquired, one can vary the contributions from these different factors and can thereby produce images that look quite dissimilar...this means that in order to read an NMR image, one needs to know not only that it is an NMR image, but also something about how it was produced.[16]

---

[16]Hill and Hinshaw 1985. Hinshaw was one of the developers of MRI devices.

Thus far I have fallen in with the usual talk of objects as the things that are observed. We now have to abandon that terminology, at least as one representing a primary ontology. We all know, although we do not always remember, that the opposition between observable and theoretical entities is a category mistake, and that the proper division is between observable and unobservable entities, on the one hand, and theoretical versus nontheoretical terms, on the other. But what are the entities on either side of the divide between the observable and the unobservable? Disputes about what is real frequently turn on objects or kinds of objects: Do prions or cartels really exist? Is there a planet beyond Neptune? What evidence was there in the late nineteenth century for the existence of atoms? Do emotions exist separately from their associated brain states? And so on. Stock examples can be misleading, for issues in realism also concern the existence of properties: How do we know that quantum spin exists? Is there such a property as alienation in society? Is covalent bonding a specifically chemical property?

In fact, properties are primary, both metaphysically and epistemologically. In the case of epistemology this is because, despite the ordinary talk about objects, it is properties and not objects that we observe or detect. Different parts of the electromagnetic spectrum give us knowledge about different properties possessed by galaxies, and different staining techniques in microscopy allow us to observe different aspects of a given specimen. Nuclear magnetic resonance imaging detects proton spins. A more familiar example concerns photographs taken with ordinary and infrared-sensitive film. The structure of vegetation as revealed by infrared photography is often quite different from the structure revealed by photographs using visible wavelengths, and observing these different properties provides us with different items of knowledge. What to those of us used to ordinary photographs looks like an extra window in your house turns out to be simply a bricked-up vent in the wall that is poorly insulated.

What kind of object is being observed can depend upon what property is being detected. The galaxy NGC 309 is usually classified as a classic spiral galaxy on the basis of its image in visible light, putting it into the same category as the Milky Way. But when viewed in the near infrared at a wavelength of 2.1 μm, the disk has only two arms instead of three and the central disk is close to an ellipsoid. This would put it into the class of barred spirals which in terms of stellar formation are quite different from the classic spirals; stars are born in the spiral arms in the latter, but in the central ellipsoid in the former.[17] As another example, the galaxy M83, when viewed by means of polarized radio continuum emission at a wavelength of 20 cm has a shape completely different from the spiral shape it displays within the visible realm.[18]

Once we see that it is properties which are detected, it is not "middle-sized dry goods" that serve as the gold standards of observability. Male rabbits are middle-sized objects, and the property of being male should be a typical observable, possessed as it is by the entire animal, but it is not always easy for veterinarians and breeders to determine whether a rabbit is male or female when it is young.[19] The prejudice that medium-sized objects are classic cases of observables results from conflating the size of an object—its spatial dimensions—with the ease or difficulty of detecting properties possessed by that object. Indeed, whether objects are considered to be macroscopic observables at all depends upon the specific properties by means of which they are detected. For example, the cloud structure in Venus's atmosphere can be seen in ultraviolet radiation but not in visible light. The entire debate on observables has been biased by an excessive focus on size and relative distance as surrogate measures for the degree of difficulty of observing objects. This bias can be exposed by considering cases where objects of the size usually associated with observable objects are indistinguishable from the material in which they are embedded unless we use sophisticated instruments. Venus's clouds are one example; magnetic resonance imaging (MRI) provides another. Some years ago a large brain tumor was detected in a patient with the aid of both CAT scans and MRI. Nevertheless, when the patient's skull was opened, no tumor was visible by optical inspection. A biopsy was then used to confirm the presence of the tumor, and it was removed.[20] So one has here a case of a macroscopic object that was unobservable with the naked eye yet unequivocally detected with quite sophisticated instrumentation applied to 'unobservables'.

[17]See Wolff and Yaeger 1993, p. 20, figs 1.10c and 1.10d.

[18]Ibid., p. 21, fig. 1.10e.

[19]My daughters' rabbits, both classified as female by their breeder, shared a hutch until one of them began exhibiting classic signs of pregnancy. Initially it was suggested that one of them, Victoria, would have to be renamed Victor. The veterinarian finally determined that the other, Niobe, was undergoing a phantom pregnancy, but noted that it was not uncommon for rabbits to be sexually misclassified at birth.

[20]See Sochurek 1988, p. 20, for this and other cases of the superiority of imaging over observation. This example indicates that the natural caution inherent in such procedures often requires a consilience of instrumental data to override plain human observations.

The ontological priority of properties suggests that the appropriate kind of realism to adopt is *property cluster realism* — entities are clusters of properties, types of entities are clusters of properties considered in isolation from spatiotemporal locations and other inessential properties. The discovery process for "objects" consists in specialized kinds of instruments detecting one or more of the properties that constitute what we consider to be that object with, subsequently, more and more properties constituting that entity being detected through further instrumental techniques.[21] We first discover the bare fact that something is causing the observed phenomena. As more comes to be known about the entity, we discover a few of the properties possessed by that object and then, with advances in instrumental detection, progressively more of them. Sometimes we are wrong, as we were in the case of thinking that phlogiston existed, but there we were wrong about the properties of what was causing substances that were heated to gain weight — there was no negative weight involved. The discovery of scientific entities thus involves a process rather like geographical discovery. In the latter case, first an indistinct property is seen on the horizon; then some of the properties composing the coastline are discovered and next to nothing is known about the interior properties; then more and more details are filled in as further exploration takes place. We can be wrong about some of those properties — just look at early maps of America, for example — and even wrong about the entities inferred from them, such as David Livingstone's view, disproved by Henry Morton Stanley, that the Lualaba River was an upper reach of the Nile, but we are rarely wrong about all of them. Even when we are, in the case of scientific properties we can always retreat to the bare existence claim that there is some property which is causing the observed effects.

2.6 Bridgman Revisited

Percy Bridgman suggested in his seminal work on operationalism (Bridgman, 1927) that different kinds of measurement gave rise to different concepts. For him, the meaning of a concept was given in terms of the set of operations used to measure the associated quantity. Thus, measuring the distance between the ends of a desk with a ruler and

---

[21]Property cluster realism can be neutral about the issue of whether individuals can be identified with the property cluster or whether they require in addition a haecceity or a "primitive thisness" (see Adams 1979; Elder 2001).

measuring the distance from the Earth to the Moon using a laser gave rise to concepts of length with different meanings. We commonly call them all 'length', but strictly speaking, according to Bridgman's position, they are different concepts, $length_1$ and $length_2$.[22] Bridgman was interested in concepts, not in properties, because he held that definitions in terms of properties (for example, of absolute space) could lead to the discovery that the definiendum failed to refer to anything real, whereas operational definitions, properly formulated, would be immune to such failures. His emphasis on meanings was unfortunate because Bridgman's insights, while susceptible to criticism as an analysis of concepts, do have value when applied to properties.

There are two different aspects of measurement that Bridgman failed to clearly separate. In some cases, he is concerned with the sorts of precautions that have to be taken when measuring something in order that the measurements be accurate. Thus, if a measuring rod was used in a magnetic field, he took this to be a different kind of operation than when the measuring rod was used in the absence of such a field, and hence the two procedures gave rise to two different concepts. If we call the kinds of procedures (operations) that correct for errors affecting a given kind of measurement, such as the measurement of length with a meter stick, 'correction operations', then, viewed from the perspective of properties, the correction operations do not take us to a different property; they merely lead, if successful, to greater accuracy on the same property. In other cases, Bridgman was concerned with something quite different: qualitatively different basic operations, such as the measurement of length by a meter stick and the measurement of length by a surveyor's theodolite. Here there are essentially different kinds of measurements that leave open the possibility that they are measuring different properties.

Bridgman's original plea not to identify concepts that are grounded in different measurement techniques came under heavy attack,[23] but looked

---

[22]"In *principle* the operations by which length is measured should be *uniquely* specified. If we have more than one set of operations, we have more than one concept, and strictly there should be a separate name to correspond to each different set of operations." (Bridgman 1927, p. 10; emphasis in original). Bridgman rejects the approach to concepts in terms of defining them in terms of properties (p. 4), and adopts the approach whereby to have a concept is to know on any given occasion how to employ it correctly (p. 5). Of course, there was always an uncomfortable tension in the slogan "the concept is synonymous with the corresponding set of operations" (p. 5), for the synonymy relation holds between meanings or words, and operations are not usually thought of as having meanings.

[23]See, for example, the influential article by Hempel (1954).

at from the perspective of properties, his caution is well advised in the case of the second type of measurement we have just described. Identifying the output of different instruments applied to the same system and calibrated so as to give the same results, however long-standing the practice, can obscure deep ontological differences between the properties that give rise to those outputs. A brief look at the case of thermometers shows this quite clearly.

We are all familiar with the mercury thermometers that mother put under our tongue when we were ill. Younger readers are no doubt more familiar with a digital thermometer of some kind, used in the ear, in the mouth, or as a strip on the forehead. These measure temperature, but so do platinum resistance thermometers, constant-volume gas thermometers, platinum/platinum-rhodium thermocouples, monochromatic blackbody radiation detectors, thermistors, disappearing filament pyrometers, vapor-pressure thermometers, bimetallic thermometers, gamma-ray anisotropy thermometers, Josephson junction thermometers, dielectric constant gas thermometers, acoustic thermometers, cross-correlated noise thermometers, and good old ethyl alcohol column thermometers.[24] Are all these measuring the same thing, a property called temperature? Not obviously.

There are three important aspects of contemporary temperature measurements. First, there is the thermodynamic temperature scale, which is based upon the idea that equal increments of temperature are intervals between which a perfect heat engine working on the Carnot cycle would perform equal amounts of work. This scale is independent of any particular material substance. Thermodynamic temperature is something that we can call an *idealized quantity*. An idealized quantity is a property that occurs only in an ideal system described by some theory and never in the real world. We cannot measure it, for it does not exist, but we can achieve successive approximations to its values. Idealized quantities such as thermodynamic temperatures are thus unlike quantities such as lengths, which, assuming space to be continuous, do correspond to elements of reality but, because those elements of reality have values that can take on real numbers, cannot be measured with complete accuracy. Thermodynamic temperatures are also unlike entities such as black holes, which are unobservable by classical criteria but which, if they exist, are real rather than ideal entities.

Second, there is the ideal gas scale, based on the ideal gas laws, where real gases are used to extrapolate back to the limit of zero pressure ideal

[24]This is only a partial list. For further examples, see Schooley 1982 and J.A. Hall 1966.

gases. This scale also marks an idealized quantity, depending as it does upon the property of being an ideal gas, and its values are also independent of the actual kind of ideal gas considered. Finally, there are the myriad real thermometers that we listed, based upon the International Practical Scale of Temperature, adopted in 1927 (the same year Bridgman's book was published) and revised periodically since then. Each of these real thermometers measures some thermometric property of a substance, this property being taken to increase monotonically with temperature.

The ideal gas and the thermodynamic scales give identical results, but each kind of real thermometer has its own deviations from this standard, so that a platinum resistance thermometer degree on the associated practical scale does not exactly equal an ideal gas degree. Moreover, different thermometers of the same kind will usually give slightly different readings on the same practical scale due to impurities in the materials used. In terms of the distinction made earlier, this latter kind of difference can be taken care of by correction operations, but the first kind of difference is due to each kind of real thermometer detecting a different property, despite giving readings on a common scale. Bridgman, even given his distrust of properties, would have agreed—different thermometers measure different practical temperatures because they detect different properties.[25]

Property realism thus shows that Bridgman's caution was justified. An ordinary mercury thermometer measures the effect of temperature on the length of a column of liquid, whereas a platinum resistance thermometer measures the effect of temperature on the electrical conductivity of a substance. Correction operations need to be made in each case to ensure accuracy—the glass containing the mercury expands and the platinum will be slightly impure—but it is the difference in properties detected that makes for the difference in temperature scales.

2.7 Instruments as Property Detectors

Instruments are property detectors. This sounds odd to the philosophical ear—surely what is detected are instances of properties and not the properties themselves? Yet natural as it is to think that way in the case of human detectors, most instruments do not detect instances because they

---

[25]More precisely, there are equivalence classes of real thermometers such that within a given equivalence class the same property is detected. There will be examples of different practical temperature scales within each such equivalence class.

are not equipped to do so. Consider two replicas of an instrument, A and B, able to detect redness in a region containing two red spheres, 1 and 2. There is nothing else in the system except a light source. The spheres are known to be free from red-shifts and there is a vacuum between the spheres and the instruments. Instrument A records the redness possessed by one sphere, and instrument B records the redness possessed by the other, but you do not know which instrument is recording the properties of which sphere. You check the output of the instruments, and each is in the same state. What did A detect? It was the property of redness. True, that property was possessed by, or was a constituent of, some sphere, but A did not detect which sphere it was nor did it detect its spatiotemporal location. It is not just that it did not record which sphere it was; it was never equipped in the first place, even indirectly, to detect the instance rather than the general quality. To put it somewhat differently, it was not 'the redness of sphere 1' that was detected by either detector, as trope theorists would claim.[26] The thing that distinguishes the redness of sphere 1 from the redness of sphere 2, which is either the spatiotemporal location of sphere 1 or its haecceity, is not detected by either instrument. I am not denying that there were originally two instances of redness. There were, but the instrument detects only the property, not the instances.

A variant on this argument will reinforce the point. You are presented with two photographs of a red sphere. The photographs are, to all appearances, pixel for pixel indistinguishable. Are they successive photographs of the same red sphere, photographs of two different red spheres, or two copies from the same original? You cannot tell, because no record of the instance or instances has been made by the camera.

There is a reason why most instruments are property detectors. Even scientific instruments that are purpose-built must ordinarily be able to repeat measurements, either on different systems or on the same system at different times. There are exceptions, such as some social science questionnaires oriented toward a specific population, Leeuwenhoek's early microscopes that were designed to observe a specific specimen at the focal point of the microscope, and instruments with spatiotemporal location recorders, but most instruments are not like those. They must be able to detect the relevant property when it is part of different objects at different locations and at different times. Move the instrument from here and now to there and then, and it will perform two internally indistinguishable

[26]See Campbell 1990.

processes of detection. Human observers are unrepresentative in this regard because we are in general able to simultaneously detect both the property and the spatiotemporal location of the source, and hence to detect that it was this instance of redness rather than that one,[27] but most instruments do not have the resources to do that. Indeed, we often must rely on humans to add that information—it was the temperature of Allie at noon on April 15 that is recorded in the physician's chart, but not by the thermometer.[28] Even when modern instruments have automatic time stamps and position locators, those spatiotemporal measurements are separate from the process of property detection because the former can be uncoupled from the latter and one can be operated without the other.

Humans have an intellectual capacity for abstraction, but most instruments, including our sensory organs, can perform acts of abstraction purely physically in at least two senses of the term 'abstraction'. They pick out from the entire constellation of properties one or a few, omitting the rest, and they also omit the spatiotemporal particularity of the instance. Most instruments are not natural nominalists. Occasionally, humans can experience what it is like to be an instrument. When I was a boy, my father would send me out on Saturday nights to get the football results. These were sold by men on bicycles who would call out "Paper" as they rode along. Because this was London in the 1950s, it was often foggy. On one occasion, two rival sellers were out. I ran through the fog, hearing "Paper" coming at me first from one direction, then another, for, as you know, sound is reflected in unpredictable ways in fog. Occasionally, the two cries of "Paper" would arrive at my ears simultaneously. On those occasions, I experienced pure numerical distinctness—two sounds having no identifiable spatiotemporal locations—and was reacting just as many instruments do.

The evening ended badly. I ran in circles for fifteen minutes without finding either source. My father was not happy.

2.8 Know Thy Instruments

...what I really wanted to see, what I was looking forward to, was the PRINCETON CYCLOTRON. That must be really *something!* ...It reminded me of my lab at home. Nothing at MIT

---

[27]There is presumably an evolutionary explanation for this ability. Knowing where a sound is coming from can be crucial to survival.

[28]The time is not far off when this will no longer be true. In many intensive care units, continuous automatic monitoring and recording of vital signs is standard.

had ever reminded me of my lab at home. I suddenly realized why Princeton was getting results. They were working with the instrument. They *built* the instrument; they knew where everything was, they knew how everything worked, there was no engineer involved, except maybe he was working there too.... It was wonderful! Because they *worked* with it. They didn't have to sit in another room and push buttons![29]

Most instruments are deliberately designed to accurately reproduce calibration standards and to produce outputs that are, ultimately, directly accessible to the human senses. Some of these instruments, such as Cassegrain reflecting telescopes, fit a simple pattern—some property of the target object initiates a causal process that is successively transformed by the instrument until the process eventually impinges on our sensory apparatus—and the geometrical optics used to understand the operation of the telescope are robust enough to render irrelevant for the user much of the detailed theory about those causal processes. One does not need to know anything about quantum optics or even much about how the telescope itself works in order to successfully use simple forms of such instruments. Historically, this was fortunate, because for many instruments, either the theory that connects the input and output was not known in detail or the one that was used was false. When Newton invented the reflecting telescope, he used the corpuscular theory of light to represent the connections between the input and the output of the instrument. The fact that the output is a faithful representation of the input was not affected by the use of this partially false theory.

Appealing to off-the-shelf examples of instruments can be misleading, for they are the finished product of a long process of testing, refinement, and adjustment. Just as the published output of scientific research can provide an overly simplified image of the crafting process that resulted in the journal article, so too can an excessive preoccupation with the kinds of simple instruments that, as examples, have tended to shape the philosophical debates on observation. Basic refracting telescopes are not representative of many other detection devices because, to use these simple instruments effectively, you now need to know very little about how they work. As with standard bench microscopes, they are designed to be so reliable that raw novices can effectively use them on suitable targets. Much more needs to be known by the user about how the instrument

[29]Feynman 1985, p. 62.

works when using research-level instruments. New research instruments malfunction, they produce spurious data, and they often need massaging to produce any data at all.

This, of course, was true of refracting telescopes when they were first introduced. We now naturally think of such telescopes as simply being made, bought, and looked through, but getting the focal length right, correcting for spherical aberration in both the objective lens and the eyepiece, adjusting the magnification for the available aperture, and so on takes a great deal of care, as anyone who has tried to construct a telescope knows and the early history of the instrument amply illustrates this. Galileo's telescopic observations of the Moon's surface, of the phases of Venus, of the moons of Jupiter, and of other previously unobserved phenomena are rightly credited with shifting the balance of evidence toward Copernican theories of the universe.[30] But, needless to say, Galileo's instruments were not precision instruments purchased from a glossy catalog and a skeptical attitude toward their outputs was amply justified at the time.[31]

As Albert van Helden notes,[32] for the first twenty-five years after their invention, telescopes were usually used for terrestrial, primarily military, observations. One reason for this was the poor quality of the images obtained on small, bright, astronomical objects: "... the best telescopes available in Holland usually show the disk of Jupiter hairy and poorly defined, whence the Jovial stars in its vicinity are not perceived well."[33] These problems would not have been disabling for terrestrial use, but for the point sources of light that are common in astronomical observations, they create serious difficulties. The story of how such telescopes were evaluated and calibrated using terrestrial sources is an instructive and entertaining historical episode that nicely illustrates the potential traps which existed in the use of an early overlap argument—for example, pages from printed texts were initially used as calibration targets to assess accuracy and resolution until it was realized that Dante's words were familiar enough to the calibrators that they could guess what lay behind

[30]The example of Galileo's telescopic observations has been discussed, for a somewhat different purpose, in Kitcher 2001.

[31]For typical responses to such skeptics, see Galileo's polemical replies to Lothario Sarsi in *The Assayer* (Galilei 1623).

[32]Van Helden 1974, p. 43.

[33]Letter of Martin Hortensius to Galileo, January 26, 1637, in Galileo, *Opere*, vol. 17, p. 19; cited in Van Helden 1974, fn. 28.

their fuzzy images. The reader is referred to Van Helden's excellent book for the details.[34]

Galileo's own observations were persuasive because he seems to have used his basic knowledge of optics to render his telescopes less prone to the deficiencies of other contemporary instruments. Although low-powered objective lenses suitable for 3x telescopes could be obtained from spectacle makers (and such low-powered devices do not reveal the crucial evidence obtained by Galileo), for his higher-powered telescopes he had to grind the lenses himself to a satisfactory standard: "The real reason that Galileo's were for a long time the only telescopes adequate for celestial observations was probably that he concentrated on the grinding of short-focus concave eyepieces. The production of such lenses entailed considerable technical difficulties."[35] One of Galileo's principal contributions was to add aperture stops to his telescopes in order to reduce problems associated with spherical aberration:

> In order to move from a three power to an eight power instrument in a very short time—a move that Dutch and French makers had not made in several months—Galileo probably did apply his knowledge of optics. If he did not, he certainly had extraordinary luck in improving the instrument to eight power, to say nothing of incredible luck about the end of the year in moving on to a thirty power telescope, which he applied to the heavens. Others were unable to produce an equivalent instrument for a very long time afterward.[36]

Much fun has been made of those who refused to believe what we think they saw through early telescopes, and this is often used as evidence to suggest that even empirical data will not convince the theoretically prejudiced. But in many cases such skepticism was justified because it was based on what was actually seen. Here is one early account:

> On the 22nd of October of last year [i.e., 1611] ... Saturn crossed the meridian. His body showed itself in three distinct scintillations through the smaller rod i.e. telescope. Through the larger rod, on the other hand, he was perceived split into four fiery balls. Through the uncovered rod, Saturn was seen to be one long single star. Therefore it appears to be the case that two,

---

[34]Van Helden 1977.
[35]Drake 1970, p. 154.
[36]Ibid., p. 142.

three, or four [*sic*] companion stars have been detected about Saturn with the aid of the new sight. But from this it does not follow that two or three servants have been assigned to Saturn by Nature.... How this fantasy can arise is evident from the above.[37]

Such uncertainty about facts as basic as how many moons, if any, orbited Saturn were a legitimate cause for concerns about the evidence. No doubt some traditionalists simply refused to admit defeat, but for others the evidence was, literally, unclear, and when historical cases such as this are put forward as justification for the claim that which theory you hold affects what you perceive and can even make you see something that is not there, or can prevent you from seeing something that is there, some caution is in order. There is no doubt that in certain cases, theoretical commitments can distort our interpretation of the data, as anyone who has spent time with conspiracy theorists will acknowledge. But as a universal claim, one that has real philosophical bite, rather than as a practitioner's warning, the evidence is unpersuasive. For example, it is sometimes said that prior to William Harvey's discovery of the circulation of the blood, anatomists "saw" holes in the septum dividing the heart because their Galenic textbooks told them they were there. But some cadavers do have holes in the septum — the famous holes-in-the-heart babies, for example — and it is not unreasonable to think that some students were presented with just that kind of cadaver. This explains, among other things, the claim that "Some anatomists, however, still maintained that the passages in the septum were easy to find in very young hearts, though concealed in the adult body."[38] There is no theory-laden observation here, simply an uncommon and sadly fatal fact.

Applied to contemporary research instruments, the point is not that knowledge is required to interpret the output of the instrument — that is an important but different issue, long recognized in the extensive training through which radiographers must go — but that knowledge is required of how sophisticated instruments work in order to reduce the incidence of artifacts until those instruments have been refined to the point where their stability can be ensured and a standard list of malfunctions is available. This issue of whether we need to understand how detectors work has a long history in the epistemology of the human senses. Do we need to know how the perceptual processes that result in human observations

---

[37]Christmann 1612, p. 41; quoted in Van Helden 1974, p. 52.
[38]A.R. Hall 1983, p. 157.

work, or can we take at least some of our perceptual data at 'face value,' as more or less veridical? Naturalists usually subscribe to the former view; more traditional epistemologists, to the latter. The question for us, committed as we are to the superiority of instruments over humans, is this: Does the need to know how the instrument works undermine arguments for the reality of the properties it detects? To make this question concrete, consider an influential argument in this area.

In Ian Hacking's article "Do We See Through a Microscope?"[39] we find a characteristically pithy set of arguments for realism. His essay is full of intriguing suggestions, not all of them fully explored, but the principal lines are fairly clear. Hacking is concerned, as his title shows, with the issue of whether we see through (or with) a microscope. But that concern concedes a little too much to the traditional preoccupation with human vision. We can instead focus on Hacking's other interest—whether the use of microscopes gives us reasons for holding a certain kind of selective realism. Hacking is an entity realist and not a theoretical realist; he holds that in certain circumstances we are justified in asserting that specific kinds of entities exist even though we may not be able to use as the basis for those assertions the truth of theories about the entities. Hacking has three core arguments for microscopic reality, which we take from a later version of his essay.

a. *The multiple independent access argument.* It is sometimes possible to observe the same structure with the aid of microscopes that use different, independent, physical processes, such as ordinary optical microscopes, fluorescent microscopes, interference microscopes, polarizing microscopes, and so on. It would be incredible to assert that there was no common physical structure that was giving rise to these common observations from different instruments: "If the same structure can be discerned using many of these different aspects of light waves, we cannot seriously suppose that the structure is an artefact of all the different physical systems."[40]

b. *The argument of the grid.* We can take a macroscopic grid, the structure of which is known, and photographically shrink it. When the microscopic grid is viewed through various microscopes, the same structure that the macroscopic grid possessed is exactly

---

[39]Hacking 1981.
[40]Hacking 1983, p. 204.

reproduced. This provides further evidence that these microscopes are giving us veridical representations.

c. *The 'don't peer, interfere' argument.* To quote Hacking, "One needs theory to make a microscope. You do not need theory to use one.... Practice—and I mean in general doing, not looking—creates the ability to distinguish between visible artifacts of the preparation or the instrument, and the real structure that is seen with the microscope. This practical ability breeds conviction."[41]

Hacking is quite clear that none of these arguments taken individually provides an irrefutable argument for realism. Our conviction that an entity exists is the product of many influences, the combination of which inclines us to accept that entity's reality. It is the first two of his arguments that are of primary interest for us, although the third also provides us with some insight into what we need to know in order to justify instrumentally based arguments for realism. Whether these arguments succeed is contingent on the case involved, and we can use them as a point of departure for a different kind of realism than the one that Hacking advocates. Here is how it goes.

We have seen that what instruments detect are properties. Each of the different microscopes cited by Hacking in the multiple independent access argument is detecting a different property. And indeed Hacking does not claim that what is real is an object. His realism is about structures.[42] He does not say what a structure is, but we can suppose that it is a spatial arrangement of properties. Hacking's multiple independent access argument will therefore succeed just in case the various modes of access, which detect different properties, all happen to produce an invariant spatial array of whatever properties the various microscopes are designed to detect. Do different properties always, or even often, present the same structures to us? No. The case of the galaxies described in section 2.5 shows that sameness of structure resulting from detection of different properties, supposedly from the same thing, often fails dramatically. This does not refute Hacking's first argument, which suggests only a sufficient condition for reality, but it does show how dependent upon contingent facts is the realism about structures that the multiple independent access argument supports.

The second of Hacking's arguments, the grid argument, is also considerably weaker than it appears. The argument is designed to show that

41 Ibid., p. 191.
42 Ibid., p. 204.

a direct comparison with known structure-preserving devices can circumvent the need to know how the instrument works. Recall the argument: A macroscopic numbered grid is photographically reduced to microscopic size, and the fact that through a given microscope one sees a faithful image of the original grid is taken as evidence that the microscope produces veridical results. The conclusion is not just that the microscope is presenting us with evidence that there is something real giving rise to what we see, but also that the instrument is giving an accurate representation of that object.

This argument, we immediately see, relies on a form of the Dilution Argument. The photographic reduction process moves an object across the dividing line between the observed and the unobserved and the microscope moves it back. As we saw, to successfully use that argument, some justification must be given that the entity which crosses the boundary retains its identity in doing so. And an appeal to preservation of a grid structure is suspect as a justification because the bench microscopes Hacking used would have been calibrated to produce an output that matched standard criteria for optical instruments. It was already known that the microscopes were free of such defects as spherical distortion because they had been specifically manufactured so that they were not subject to flaws that would have misrepresented the grid. Any microscope that failed these tests would have been rejected in the factory and never would have reached the laboratory. Reproducing the calibration standard, or something sufficiently similar to it, cannot be used as evidence for the reality of that standard because the instrument has been deliberately configured to accurately present phenomena at that level. Moreover, it is not true that even bench microscopes faithfully reproduce spatial structure, because some spatially invert the image. We discount that as unimportant, but on what grounds? Not that we have direct access to the object of magnification, as we do with certain telescopes, but on the basis of well-established theory about lenses. Without that, the hypothesis that the photographic reduction process inverts images would not be excluded.

The point here is that the theoretical knowledge which is embedded in the construction of the microscope and which has made the instrument a stable and reliable producer of veridical data has, in the case of the grid, made the use of theoretical corrections unnecessary. As with many instruments, one already has a guarantee that on standard sources they either correctly represent objects similar to the calibration standard or that deviations from accuracy are known. (Recall in contrast the Hubble telescope fiasco, where the calibration instrument was itself maladjusted.) But if we consider instruments that are not catalog models, that is, instruments which are unlike the bench microscopes in that they have not been deliberately manufactured to be student-proof, the theory of how the instrument operates is routinely needed to protect the user against skeptical objections resulting from ways in which the instrument can produce misleading outputs.

Such precautions are commonplace. Consider our earlier example of magnetic resonance imaging (MRI). An external magnetic field lines up the spins of the protons in hydrogen nuclei. A brief radio signal is used to distort that uniformity, and as the hydrogen atoms return to the uniform lineup, they emit characteristic frequencies of their own. Computers process this information to form an image suitable for display to the human visual apparatus, this image representing various densities of hydrogen and its interaction with surrounding tissue. MRIs operate differently with respect to water than do the human senses of vision and touch in that rather than detecting water by means of its surface features as we do in everyday life, the instrument uses part of the inner structure of water to detect its presence. One result of this method of detection is that it enables one to avoid imaging spatially intermediate entities even though they are in the 'direct line of vision' between the MRI scanner and the target, one of the many items of knowledge necessary to effectively use the device.

A story suggests another reason why you need to know how MRI devices work in order to use them effectively. The first MRI scanner that was put into use at Cornell Medical Center consistently produced seriously distorted images. After a number of attempts to explain what was going wrong, it was finally discovered that a radio station was broadcasting on the same frequency that the scanner was using and thus interfering with the operation of the device.[43] Knowing how instruments work is important mostly when they go wrong or give unexpected outputs, particularly when operating outside the domain on which they overlap with existing instruments.

To take a simpler example, when installing an internal/external thermometer, one has to know a little bit about how the thermometer works in order to obtain a reliable reading. Installing it on a north wall out of the sun (for accurate air temperatures), asking whether it will work horizontally as

[43]See Sochurek 1988, p. 10.

well as vertically (it will), and realizing that the uninsulated mud room will still be warmer than the outside air are just three factors that are important in setting up this simple instrument to ensure accuracy.

The technique of quantitative electron microprobe compositional mapping provides a final, more sophisticated, illustration of these points. With these instruments, a spectrometer (either a wavelength or an energy spectrometer) measures the count rate for each element within a section of a grid, compares it against the count rate for a pure sample of the element, and constructs a matrix of intensities for the sample. There are two principal instrumental errors involved with compositional mapping— defocusing of the wavelength spectrometer and decollination of the energy spectrometer, and variation in the background radiation that depends on the composition of the material. The techniques needed to correct for these errors require significant understanding of how the spectrometer works. The important philosophical point here is that once we understand how the apparatus operates, we have explicit control over the parts of the observation process that are governed by that theory. We know why wavelengths that are off the optical axis produce distortions in the intensity distribution, and thus how to correct for it. Because that working knowledge is explicit, it poses no threat to the security of the evidence that the output gives us—in fact it strengthens it by assuring us that one source of distortion has been removed. The theory dependence of observation here is not a defect, it is a positive virtue.

Often, it is not the direct image that provides the best information about the subject matter, but an image that has been deliberately manipulated, and manipulation of the images in compositional mapping provides additional information. If one takes compositional maps of the elements in a material, and represents them with primary colors, then superimposing those maps will immediately give a secondary color image that represents the distribution of the joint presence of the elements. Other commonly used techniques that distort the original are color enhancement, which involves taking all the pixels of a given color and enhancing them; contrast enhancement, which takes all pixels with a brightness level below a given threshold and decreasing their brightness while simultaneously increasing the brightness of those above the threshold; and edge detection, which looks for sharp changes in color in an image and then blackens out the surrounding areas, giving a 'bare bones' geometrical structure. Again, knowing how these images are produced is essential both to interpreting them and to avoiding error.

Methodologically, this reliance on knowledge of how the instrument works is important, because the significance of Hacking's approach lies in its forthright rejection of the possibility of arguing for realism through theories. If we did indeed have theory-independent access to the world, it would be a signal advantage for him. The principal advantage in knowing how the instrument works is that without that knowledge, we have to rely on the correlations between the input and the output remaining invariant. This reliance is justified as long as the instrument continues to operate without error. But it is exactly when conditions vary and the invariance breaks down that we need to know whether we have gone outside the instrument's domain of application and, if so, how to correct for it. In most cases you cannot know that without knowing how the instrument works, but with that knowledge, you can often adjust for those changes in operating conditions.

This gives us some insight into why there has been such an emphasis on the unaided senses as a source of reliable information. It is in part because we are so familiar with the circumstances under which they fail us. We ordinarily assume that the use of our senses in situations outside ideal contexts is relatively unproblematical, but this is based on our quite refined sense of the conditions under which the evidence of our senses becomes unreliable by loss of accuracy, by degraded precision, or by producing artifacts.

To conclude this discussion of Hacking's argument, he initially takes the position adopted here, at least with respect to the seeing question: "To understand whether she was seeing, or whether one sees through the microscope, one needs to know quite a lot about the tools."[44] But contrast this with his later claim, a claim that is crucial for his overall conclusion that one can circumvent the dependence of observation on theory: "One needs theory to make a microscope. You do not need theory to use one."[45] As we have seen, this is sometimes correct and sometimes not. You do, as we have seen, often need to know how the instrument works in order to use it effectively, but this knowledge is not always based on theory or, even when it is, upon a correct theory. In the COBE (Cosmic Background Explorer) satellite that was used to explore the 3°K background radiation, there was an onboard 3°K reference source with which the observed radiation was compared. Because a direct comparison was made between

[44]Hacking 1981, p. 307.
[45]Ibid., p. 309.

the calibrator and the source, little theory was required to justify the results. What was important was knowing the ways in which the instrument could go wrong, its limitations, and what alternative explanations there could be for the observed data.

There exists a wonderful photograph of Ernest Rutherford's laboratory that nicely illustrates why theoretical knowledge can be unnecessary.[46] Rutherford's voice was so loud that researchers were forced to hang a sign reading TALK SOFTLY PLEASE to prevent sound waves affecting their experiments. No fancy theory of sound was needed—the experimenters simply knew from experience that when Rutherford was around, their experiments went awry. Because it is not always easy to distinguish theoretically based knowledge from practical knowledge in any given field, it is preferable to put these points in terms of the need for *subject-specific knowledge* rather than theoretical knowledge in using these instruments. The more we know about how an instrument works, the more justification we have about the degree of faithfulness with which the data represent the object observed.

2.9 Properties

The fact that many instruments are property detectors in the sense detailed earlier follows from a striking feature of causal properties, one that perhaps more than any other determines their nature. It is important enough to have its own name: *Property Localization Principle*: Each instance of a causal property possesses the entire causal force of that property.

This fact suggests that we should adopt a certain kind of ontology for causal relations, including those which we rely upon in measuring instruments. First, it is properties and not objects that cause effects. It was the iron's property of *being hot* that caused my hand to be burned. It was not the iron that did it, because the iron could have been there but cold. Second, although a property must be spatiotemporally located to be a cause or an effect, it is not the instance of the property that does the causing but, once again, the property itself. Was it *this iron's* being hot that did it; that is, the possession of hotness by the iron rather than simply the hotness? Again no, but less straightforwardly. It was not the iron's being hot, because had something else with the same temperature been there instead, the burn would have occurred as it did. As it did, but likely

---

[46]See Oxbury 1985, p. 295.

not exactly as it did, for it is ordinarily a combination of properties that produces an effect rather than a single property. It is these clusters of features, which are often characteristic of particular types of objects, that lead us to elliptically attribute causation to the objects. Superheated water of the same temperature as the iron would have produced a different kind of burn because some other properties associated with water but not with metal—wetting and lower conductivity, for example—would have produced a somewhat different effect. But even when we identify an object with a spatiotemporally located cluster of properties, it will rarely, if ever, be the entire cluster that is involved with the production of the effect.

Was it the hotness there and then that did it? That depends upon what 'it' is. Science is usually interested in effect types—that is, the property or collection of properties that constitutes an effect, and in general, where and when a cause acts is not relevant. Of course, for pragmatic reasons, we might be interested in what caused the burn where and when it occurred, but qua burn, it could have happened elsewhere and elsewhen and in exactly the same way.

Spatiotemporal relations between two properties can make a causal difference, but not the spatiotemporal locations themselves. Aristotelians thought otherwise—the natural place in the universe for an element could cause a piece of that element to move toward it—but spatiotemporal particulars are no longer considered to have such a force. This fact lies behind two core features of science. The replicability of scientific results rests upon the fact that exactly the same result can occur in more than one location. Replicability is primarily an epistemological precaution to guard against fraud and error but, as we have seen with instrument detection, its very possibility rests upon rejecting property instance specificity. The second feature is the universality requirement on scientific laws; more specifically, that no reference to spatiotemporal particulars should be included in laws. That feature, which is sometimes motivated by an appeal to testability but is more often grounded in an appeal to our sense of what constitutes a law, can be explained in terms of the property localization principle. Many laws give us relations between properties, and because of the property localization principle, references to spatiotemporal features not only are unnecessary but also should be banned because they misleadingly restrict the scope of the law.

It thus was the property, instantiated, but not necessarily by the iron, that did it. It might be said that rather than some specific object's being hot, it was simply something's being hot that did it. This would essentially concede the point made here. Because it is irrelevant what object makes that quantification true, while it is crucial to have the right property, it is the property and not the object that is involved in the causing.

How did the property do it? Not, as some nominalists tell us, by virtue of hotness being scattered around the universe.[47] The hotness present on the surface of some star in a galaxy millions of light-years away is irrelevant to what caused my hand to hurt—take away that remote hotness, or any other instance, or every other instance, and nothing changes about the ability of the hotness right here and now to cause me pain. Suppose, more strongly, that a property is the set of all its actual and possible instances. (All the instances of what? The property? A queer saying, but we'll let that pass here.) Such is the way that David Lewis[48] construed them. Then a similar difficulty can be raised for that conception. Remove a remote or possible instance and you have a different set, hence a different property. So the removal of the remote instance would mean that it was not an instance of hotness, but of something else that caused my burn. Something remarkably similar, but still not the same property. Of course, one might say that you cannot take away an instance of hotness because on Lewis's view the set of possible worlds is what it is, and 'taking away an instance of hotness' is impossible. All that you mean is that you are considering a world very similar to ours, but without that instance, and indexically construing that world as the actual one. This is true, and although Lewis notes that by using counterparts, you can maintain the contingency of, say, this iron's being hot, it is still true that the set of hot things is not contingent. It necessarily has the members that it does—in every possible world, the extension of hotness is the same set. So Lewis's account is immune to this objection. However, it retains the curious feature that what makes the (causal) property the one it is, and not something perhaps entirely different, is its nonlocalized extension. So in a world where there is but one instance of hotness, what makes it hotness, rather than some other property that also happens to be uniquely instanced in that world, is

---

[47]Taken as a general truth about properties, this is false. But some have held it true for certain properties. Quine, for example, writes: "Earlier we represented this category by the example 'red', and found this example to admit of treatment as an ordinary spatio-temporally extended particular on a par with the Caÿster. Red was the largest red thing in the universe—the scattered total thing whose parts are all the red things. . . . So, the theory of universals as concrete, which happened to work for red, breaks down in general . . . triangle, square, and other universals were swept in on a faulty analogy with red and its ilk." Quine 1961b, pp. 72–73.

[48]Lewis 1986, pp. 50ff.

a set of spatiotemporal particulars located at some other worlds, themselves utterly causally isolated from the original. Whatever advantages the 'properties as sets of instances' view has for other purposes, such as extensional formal semantics for various logics, it seems ill adapted for putting properties to causal work.

This line of argument in favor of property causation is similar to one that can be brought against regularity accounts of causation. Hume and his descendants require that for X to cause Y, there must be a regularity of X-type events followed by Y-type events. You might make a case for this view on epistemological grounds, but metaphysically it is unpersuasive. I discover my bicycle tire is flat. Curses! How did that happen? I inspect the tire and there it is, the proximate cause, having a hole (in the tire). Now suppose that every other instance of a tire having a hole in it, followed by the tire's being flat, were absent from the world. (If you dislike thought experiments, consider the first ever case of a pneumatic tire receiving a puncture.)[49] What possible difference could removing the other instances of the regularity make to the causal efficacy of this instance of having a hole bringing about the flatness in the tire? None, because the entire causal force of having a hole is present at the exact location where that property is instanced, just as the hotness exerted its effect at the exact location where my hand touched the iron, and nowhere else. It makes no difference, by the way, whether the bearers of the properties are described by mass terms or by count nouns. Take a swimming pool. The left side is filled with stuff that wholly and completely has the properties of water. So is the right side. And so, of course, is the whole.[50]

---

[49]The actual case is weaker than the hypothetical case because the latter can be applied to any instance, not just the first. Moreover, the actual case is vulnerable to objections from those who hold that the future is already in place. Thus the latter is preferable. Those who hold that laws supervene upon matters of particular fact are subject to this counterexample if, like regularity theorists, they require lawlike regularities for causal relations in the sense of 'No regularities, no causes'. By the way, I mean 'thought experiment' here, and not 'different possible world'. What's the difference? In the thought experiment, you know that in the real world, some holes in tires could have been prevented by, for example, using real tire irons rather than spoons to put in a new inner tube. Then, as some thought experiments do, just extrapolate from that.

[50]Of course this point cannot be iterated arbitrarily often unless you believe, as many apparently do nowadays, that a single molecule of H_2O is itself water. I find this a curious view, for individual molecules of water do not have all the properties of water, such as liquidity, despite the popular dogma of their identity. Some properties of water are a result of features that occur only in very large ensembles of interacting molecules. You can't stipulate *that* away.

To return to the striking claim that introduced this section, we can, I think, make sense of a claim considered by David Armstrong[51] that a property is wholly and completely present at each of its instances, although not, I think, in a way that Armstrong himself intended. This assertion seems clearly false because it contradicts the claim that nothing can simultaneously be wholly and completely present at two distinct spatial locations, which many have taken to be necessarily true. One naturally thinks of this claim in spatial terms: The whole of a property cannot be present in this bit of space and all of it simultaneously be wholly over there. Perhaps, but because one role that properties play is that of causal agents, if we think of this assertion as meaning that a property can wholly and completely exert its entire causal effect at location **a** and simultaneously wholly and completely exert its entire causal effect at a different location **b**, then it is (obviously and contingently) true.

2.10 Epistemic Security

Some final words are in order. One line of argument for liberating empiricism beyond the realm of what is directly accessible to the unaided senses hinges on the contingency of our sensory equipment. Our visual apparatus, it is said, could have evolved to be more acute than it is, allowing us to observe ninth-magnitude stars with the naked eye. Ordinarily, human observers require an optical telescope to observe stars of brightness fainter than the sixth magnitude but, so this argument goes, that is an accidental feature of our evolutionary development and thus it is of no consequence for empiricist epistemology. And similarly for extensions to microscopic perceivability, to the ability to perceive electromagnetic radiation in parts of the spectrum outside the visible range, and even to the augmentation of our senses.

This contingency should be taken seriously, but before the development of extrapolation and augmentation devices, the force of the argument was severely limited. Had such instruments never been invented, the traditional empiricist could, justifiably, have maintained that the possibilities involved were idle and that the only kind of epistemic agent of interest to us was one of our own. That response, given the arguments of this chapter, is now unappealing. The contingencies of biological

[51]Armstrong 1989, p. 98.

evolution have been radically enhanced by the contingencies of technological evolution, and as a result many formerly idle possibilities are hard
at work. Entification now begins at charm's length.

A similar point can be made about the role played by enhancers of
our native computational abilities. The fact that we often need instrumental help for our calculations is also a contingent fact. The proofs that
led to the classification of finite simple groups originally ran to a total of
15,000 pages, and part of the classification of the sporadic groups was
completed with the aid of computers. If, contrary to fact, we were equipped
with much faster brains and much longer life spans, this computational
help would have been unnecessary. The empirical and the computational
cases differ in that the contingency in the empirical case is a double
contingency—we could have had different abilities and the world could
have been different, whereas in the mathematical case there is only a
single contingency because the mathematical facts could not have been
other than they are, but the overall point remains the same.

The traditional empiricists' desire to stay close to their natural abilities and their aversion to error were part of their response to skepticism,
the suggestion that for all we know, the world might be radically different
from the way we believe it to be, appearances to the contrary. Many such
suggestions are of great importance to fundamental epistemology, but
those concerns of traditional epistemology usually are conceptually prior
to the concerns of philosophy of science and it is a mistake to believe that
scientific investigations can solve them. In particular, radical global skepticism has to be put aside when we consider what science can tell us.
What is characteristic about global skepticism is that the alternative explanations are compatible with all actual and nomologically possible
evidence, and so it is by its nature incapable of refutation by science.[52]
Science does not, should not, and could not worry about fantasies which
suggest that when we inspect the results of a bioassay, an evil demon is
systematically making us misread the results or that when we are using a
mass spectrometer, we are all dreaming.

The situation is radically different when we move from global to local
skepticism because one of the characteristic differences between science

---

[52]Nomologically possible, not logically possible, because it presumably is logically
possible that we could come to empirically discover that deceptive demons inhabited our
world, or that there were intrinsic, qualitative differences between dream states and waking
states from evidence that was purely internal to one or the other of the two kinds of state.

and philosophy is that the former always uses a selective and not a global form of skepticism.[53] Yet even classical empiricism cannot, except in austere phenomenalist forms, be a successful response to global skepticism. It is instead a way of responding to local skepticism. Doubtful that there is an iguana in the next room? Go and see for yourself. Suspicious that the Bordeaux might have spoiled? Taste it and decide. This sort of response to local skepticism is unobjectionable but it is severely limited in its scope. Even within its restricted domain, the traditional empiricists' response to local skepticism requires us to segregate possible explanations for the phenomena into those which we must take seriously and those which are mere philosophical fantasies. The waiter might suggest that the wine is fine and that you have been hypnotized to believe that all 1969 Lafite has the aroma of Limberger cheese, but when there is no evidence to support this, it would be correct to reject it as a credible alternative explanation for what you are experiencing. As a logical possibility, it is scientifically idle. It is the responsibility of a legitimate scientific empiricism to allow science to help us decide which alternative explanations for the instrumental data should be taken seriously and which should not.

Once the superior accuracy, precision, and resolution of many instruments has been admitted, the reconstruction of science on the basis of sensory experience is clearly a misguided enterprise. Indeed, one consequence of the view defended here is that the logical empiricists had their epistemological priorities reversed and that progress against local skepticism is more often made by using scientific evidence than by using ordinary sense perceptions.

The empiricists' attempt to minimize the possibility of error has, as a consequence of the quest for certainty, historically been asymmetric, focusing primarily on the chance of wrongly believing a proposition to be true when it is not. But there is another kind of error, that of wrongly

---

[53]Part of a proper education should be discovering the appropriate balance between warranted and unwarranted skepticism in a given area. What is warranted varies between areas; more skepticism is warranted with respect to postmodern studies than with Galois theory, for example. In everyday life we regularly employ selective skepticism; unselective skepticism can lead to paranoid delusions and an inability to learn. Contemporary theories of contextualism hinge on the issue of selective skepticism: See, e.g., DeRose 1995; for an often unacknowledged anticipation of that view, see Cargile 1972. Those readers who are uneasy at the thought of delegating the rejection of global skepticism to epistemology proper should simply read what follows as describing how to acquire the analogue of knowledge when cordoning off their favorite skeptical scenario. For a disastrous instance of credulity on the part of a physicist, see Taylor 1975.

believing a proposition to be false when it is actually true.[54] The two types of epistemic error correspond to, respectively, Type II and Type I errors in statistics—accepting the null hypothesis when it is false and rejecting it when it is true. A common parallel is with judgments in courts of law—a Type I error corresponds to convicting an innocent man (wrongly rejecting the true hypothesis of innocence) and a Type II error corresponds to allowing a guilty man to go free (wrongly accepting the proposition of innocence as true). In general, it is impossible to simultaneously optimize these two types of error, and there are many situations in which the Type II error—the one traditional empiricists strive to avoid—is of lesser importance than the Type I error. The excessive caution ingrained in traditional empiricism thus has its downside. Forgoing the possibility of discovering new truths because of a reluctance to be wrong is not always epistemically optimal.

Through all this it has been the safety of the inference that is at stake, and not a permanent, intrinsic difference between the observability of one thing and the nonobservability of something else. There is no reason for scientific empiricism to be so constrained by the highly contingent and limited resources of the unaided human sensory apparatus, and it can be extended without overly weakening its traditional orientation. Epistemology need not be confined to the study of human knowledge obtainable by us as we are presently constituted. None of the positions traditionally falling under the category of empiricism can properly capture the knowledge produced by many new methods in the sciences. It is within the vast territory shunned by these positions that most scientific knowledge is located, and I take the view that instrumentally and computationally guided understanding is sui generis, not to be reduced to something more elemental.

---

[54]There are other kinds, such as refusing to believe a proposition true or false under the mistaken belief that it has no truth value and, conversely, attributing a truth value to it when it actually has none. I shall ignore these possibilities here.

3

Computational Science

3.1 The Rise of Computational Science

> It is now possible to assign a homework problem in computational
> fluid dynamics, the solution of which would have represented a
> major breakthrough or could have formed the basis of a Ph.D.
> dissertation in the 1950s or 1960s.[1]

Philosophers have paid a great deal of attention to how computers are used
in modeling cognitive capacities in humans and to their use in the construction of intelligent artifacts. This emphasis tends to obscure the fact that
almost all of the high-level computing power in science is deployed in
what might appear to be a much less exciting activity—solving equations.
In terms of the computational resources devoted to various areas of science, the largest supercomputers have always been located in nuclear
weapons research facilities, with weather and climate modeling a more
recent beneficiary.[2] This situation reflects the historical origins of modern
computing. Although Alan Turing's work in the period 1940–1945 with
the ENIGMA group on cryptanalysis, using the Colossus electronic
computer,[3] and his 1950 paper on artificial intelligence[4] have justifiably
attracted much attention, as has his earlier work on the foundations of
recursion theory, most of the early electronic computers in Britain and
the United States were devices built to numerically attack mathematical

---

[1]Tannehill et al. 1997, p. 5.

[2]Because of the pace of development in supercomputing, as well as differences in
methods for measuring performance, it would be pointless to list here the current fastest
machines. Up-to-date lists can be found at www.top500.org.

[3]See Hodges 1983.

[4]Turing 1950.

problems that were hard, if not impossible, to solve nonnumerically, especially in the areas of ballistics and fluid dynamics. For example, the ENIAC computer built at the University of Pennsylvania in the period 1943–1945 was originally designed to calculate artillery shell trajectories, and one of its first uses was at the Aberdeen Proving Grounds for a one-dimensional simulation of a nuclear explosion.[5] The latter kinds of calculations were especially important in connection with the development of atomic weapons at Los Alamos.

Computational methods now play a central role in the development of many physical and life sciences. In astronomy, in physics, in quantum chemistry, in meteorology and climate studies, in geophysics, in oceanography, in population biology and ecology, in the analysis of automobile crashes, in the design of computer chips and of the next generation of supercomputers, as well as of synthetic pharmaceutical drugs, of aircraft, of cars and racing yachts, and in many other areas of applied science, computer simulations and other computational methods have become a standard part of scientific and engineering practice.[6] They are also becoming a common feature in areas such as econometrics, many areas of the social sciences, cognitive science, and computer science itself.[7]

Why should philosophers of science be interested in these new tools? Are they not simply a matter of applied science and engineering? To view them in that way would be to seriously underestimate the changes which these methods have effected in contemporary theoretical science. For just as observability is the region in which we, as humans, have extended ourselves in the realm of the senses, so computability is the region in which we have extended ourselves in the realm of mathematical representations. And not computability in principle, but computability in practice. It is the solvability of models that is the major barrier to the application of scientific theories, and if we restrict ourselves to the resources of the human mind augmented by pencil and paper, the solvability constraint bites hard, even with severe idealizations and approximations. The developments sketched above, which began in the 1940s and accelerated rapidly in the last two

---

[5]Metropolis 1993, p. 128.

[6]As early as the mid-1980s, for example, 60%–70% of journal articles in theoretical astrophysics used computational methods at some point in the research. The current holder of the world land speed record, the supersonic Thrust SSC, was designed using simulators, as are all commercial airframes.

[7]For applications in other areas, see Conte et al. 1997; Hegselman et al. 1996; and Holland 1993.

decades of the twentieth century, have given rise to a new kind of scientific method that I shall call *computational science* for short. It has a number of different components, some familiar and some not so familiar, but taken together they result in ways of doing science that do not fit the traditional trichotomy of theory, observation, and experiment. An important subinterest will be the area that generally goes under the name of computer simulation, and we shall need to carefully distinguish this area from other aspects of computational science.

Here is what a few internal commentators have had to say about these new methods:

> Science is undergoing a structural transition from two broad methodologies to three—namely from experimental and theoretical science to include the additional category of computational and information science. A comparable example of such change occurred with the development of systematic experimental science at the time of Galileo.[8]

> The central claim of this paper is that computer simulation provides (though not exclusively) a qualitatively new and different methodology for the physical sciences, and that this methodology lies somewhere intermediate between traditional theoretical physical science and its empirical methods of experimentation and observation.[9]

> For nearly four centuries, science has been progressing primarily through the application of two distinct methodologies: experiment and theory.... The development of digital computers has transformed the pursuit of science because it has given rise to a third methodology: the computational mode.[10]

These are bold claims. The situation is more complex than the quotations indicate because the introduction of computer simulation methods is not a single innovation but is a multifaceted development involving visualization, numerical experimentation, computer-assisted instrumentation, individual-based modeling, and other novelties. It would be possible to hold different, less radical, views on these developments. One view would be that the methods of computational science are simply numerical methods that have been greatly enhanced by fast digital computation devices with large

[8]*Physics Today*, 37 (May 1984): 61, quoted in Nieuwpoort 1985.
[9]Rohrlich 1991, p. 507.
[10]Kaufmann and Smarr 1993, p. 4.

memory capacity and that they introduce nothing essentially new into science. A second view would hold that these techniques do not constitute legitimate science, in part because numerical experiments, being free of direct empirical content, cannot substitute for real experiments. A third objection would argue that because the correctness of some of the computational processes underlying theoretical computational science cannot be conclusively verified, they fall short of the standards necessary for scientific methods. These objections need to be addressed, although ultimately they are based on a lack of appreciation for the ways in which computational science differs from more traditional methods.

Consider the response that computational science introduces nothing essentially new into science. This response fails to recognize the distinction between what is possible in principle and what is possible in practice. It is undeniable that in some sense of 'in principle', advanced computational methods introduce nothing methodologically new. But this level of abstraction from practice is inappropriate for evaluating scientific methods, for to claim that 'in principle' the computational methods are no different from traditional mathematical methods is, to put it mildly, disingenuous. The idea that an unaided human could 'in principle' provide a numerical solution to the equations needed to predict weather patterns over the East Coast a week from today is a complete fantasy. Billions of computational steps are involved in sophisticated computer simulations, and no pretense can be made that this is open to checking or duplication in a step-by-step fashion by a human or by teams of humans. An average human can calculate at the rate of 10^{-2} floating point operations (flops) per second. (One flop is the equivalent of adding or multiplying two thirteen-digit numbers.) Current supercomputers operate at teraflop speeds: 10^{14} times faster than humans. A teraflop machine operating for three hours can perform a calculation it would take a human the age of the universe to complete,[11] and current supercomputer speeds will seem

---

[11]This estimate is taken from Kaufmann and Smarr 1993, pp. 176–177. There persists in the literature a tendency to write as if Turing machines underlie everything that is done. Even writers who clearly know better lapse into this mythical mode: "When the computational task is to simulate a physical phenomenon, the program has traditionally been built around a partial differential equation.... For example, Maxwell's equations describe the behavior of electrical and magnetic systems; the Navier-Stokes equation describes the behavior of fluids. Although the equations themselves are quite complex, it is known that they can be solved with an extremely simple computing apparatus ... known today as a Turing machine" (Hasslacher 1993, p. 54). Hasslacher is a physicist at Los Alamos National Laboratory.

laughable a decade from now. This extrapolation of our computational abilities takes us to a region where the quantitatively different becomes the qualitatively different, for these simulations cannot be carried out in practice except in regions of computational speed far beyond the capacities of humans. Speed matters.

The restricted form of the second objection is an important one, and the dangers of using simulations devoid of contact with empirical data are indeed significant. Concerns along these lines will be addressed in section 4.7, where potential abuses of computer simulations are discussed. There is no sense in which I am advocating the view that computational science should be done using only simulated data, with the exception of when exploratory work is being carried out. But this specific kind of objection should be sharply separated from the more general claim that computational science simply does not count as a genuine kind of science. The case against this pessimistic view will have to be made over the course of this chapter and the next. For the moment, however, this can be said: For those of us who are interested in how theories are applied to nature, the most important immediate effect of using these methods of computational science is that a vastly increased number of models can be brought into significant contact with real systems, primarily by avoiding the serious limitations which our restricted set of analytic mathematical techniques imposes on us. Those constraints should not be underestimated, for once we move past the realm of highly idealized systems, the number of mathematical theories that are applicable using only analytic techniques is very small. Nor could it be said that humans have no need to duplicate the machine's computational processes because they have access to other, more efficient, ways to solve the equations. For that response, as we shall see in section 3.4, is unavailable for most models.

To fully appreciate the scope of these methods, one must be willing to set aside the logical analyses of theory structure that have been so influential in the philosophy of science, to switch attention from problems of representation to problems of computation, and to relinquish the idea that human epistemic abilities are the ultimate arbiter of scientific knowledge. Our slogans will be mathematics, not logic; computation, not representation; machines, not mentation.

In suggesting this, in no way am I endorsing the idea that logical methods, philosophical theories about representation, and, most important, thoughtful understanding and use of these methods are irrelevant or misguided. Quite the contrary. The points are, rather, that accounts of abstract theory structure cannot properly capture how theories and models are constructed, modified, and applied; that intractable representations are scientific bystanders; and that how we understand and appraise these new methods is essentially different from the understanding and evaluation of traditional theories.

The third complaint about the lack of verifiability of artificial computational processes compared to human processes involves a fallacy of composition. It is true that in many cases, any specific individual step in a complex computation can be checked for correctness with complete reliability by a human, but it of course does not follow from this that all steps in the complete computational process can be so examined seriatim with complete reliability. We all know this from trying to balance our checkbooks. There are once again analogies with the limitations of the senses. Consider a human visually inspecting mass-produced Lego blocks on an assembly line.[12] There are a dozen different standard kinds, and the inspector must check them for defects as they move past. He can slow down and stop the assembly line whenever he wants to check an individual unit in detail, but there are many thousands of blocks to be checked. By the end of the day he has to complete the task and certify that every single unit is error-free. Accuracy of the inspector on any individual part is in isolation very high, but it should be clear that this is no guarantee of the inspector's sequential accuracy over the entire output. The same point applies to complex mathematical proofs, and the preface paradox rests on our acknowledging a similar point about truth and argument.[13] With regard to the three criteria of accuracy, precision, and resolving power that we have used to evaluate instruments, artificial computational devices provide increased accuracy on almost all tasks, superior precision, and far greater resolving power when using numerical methods, in the sense that the degree of resolution attainable in the solution space with spatial and temporal grids on computers lies well beyond the reach of paper-and-pencil methods.

---

[12]Lego blocks are small plastic interlocking bricks that come in a limited number of basic shapes, sizes, and colors. From these units, structures of arbitrary complexity can be built, such as replicas of the Houses of Parliament.

[13]The preface paradox concerns a hypothetical author who asserts in the preface to his book that he is sure of each individual argument he has constructed and of each conclusion he has reached, but nevertheless cannot guarantee that the book contains no errors. It is part of the accepted wisdom of mathematics journal editors that almost all papers contain at least one error, almost always minor.

We can sum up the attitude taken here as follows. Traditional empiricism and rationalism take what is humanly accessible as being of primary interest. "In principle" epistemic possibility serves as the boundary for the former, and logical possibility accessed through the a priori for the latter. In contrast, we take what is technologically possible in practice as the primary constraint on contemporary science. The original and expanded domains for three traditional methods within science can then be represented by positions in a three-dimensional space where the axes represent what is observationally detectable, what is mathematically accessible, and what is open to experimental manipulation. As an exercise for the reader, I recommend placing the various techniques discussed in this book within this epistemological space.

3.2 Two Principles

We can begin with a point that ought to be uncontroversial, but is not emphasized enough in the literature on the philosophy of science. It is that much of the success of the modern physical sciences is due to calculation. This may seem not to be a very interesting activity, but in getting from theory to application it is almost indispensable. Perhaps because of its lack of glamour, but also for deeper structural reasons, twentieth-century philosophy of science has failed to accord proper credit to scientific calculation beyond the philosophy of mind. This is curious, for behind a great deal of physical science lies this principle: *It is the invention and deployment of tractable mathematics that drives much progress in the physical sciences.* Whenever you have a sudden increase in usable mathematics, there will be a sudden, concomitant increase in scientific progress in the area affected.[14] And what has always been true of the physical sciences is increasingly true of the other sciences. Biology, sociology, anthropology, economics, and other areas now have available, or are developing for themselves, computational mathematics that provide

[14]Empiricist philosophers of mathematics have sometimes emphasized the importance of this symbiotic relationship. Imre Lakatos is quite clear about it: "The real difficulties for the theoretical scientist arise rather from the *mathematical difficulties* of the programme than from anomalies. The greatness of the Newtonian programme comes partly from the development— by Newtonians—of classical infinitesimal analysis which was a crucial precondition for its success" (Lakatos 1970, p. 137). There are also scattered remarks in Kitcher 1984 to this effect. In the other direction, Morris Kline has frequently argued that the development of mathematics is driven by the requirements of science (see, e.g., Kline 1972, passim), although it is easy to overstate that kind of influence.

a previously unavailable dimension to their work The investigation of fitness landscapes, the modeling of cultural evolution through learning and imitation, game-theoretic models of economic agents with bounded rationality—these and many other enterprises have been made possible or greatly enhanced by modern computational resources.

There is a converse to this principle: *Most scientific models are specifically tailored to fit, and hence are constrained by, the available mathematics.* Perhaps the most extreme example of this is the use of often highly unrealistic linear models in the social sciences, where the severe limitations in nonlinear statistical methods make linear models almost inescapable, but the scientific landscape in general is shaped by computational constraints. Select any text on 'Mathematical Methods in X' and you have a good grip on what can and cannot be done in the way of applying the field of X. Mathematical models that are unsolvable have some sort of representational interest and certainly can play a significant role in advancing our understanding of an area, but for applying and testing theories they are, literally, useless.

All of this is straightforward, but a natural inclination toward abstract approaches on the part of philosophers can lead us to neglect it.[15] Applications of the two principles can be seen at both the large and the small scale. At the small scale, the invention of methods in the mid-eighteenth century to solve second-order partial differential equations led to significant advances in the field of vibrating strings.[16] On a grander scale, there is the familiar use of the epicyclic methods in Greek astronomy, developed by Apollonius, Hipparchus, and their contemporaries and systematized by Ptolemy, which formed the basis of an extraordinarily successful mathematical astronomy for more than 1700 years. It is often said that this apparatus was motivated by Platonic conceptions of the universe, but the existence of such a powerful set of methods influenced many astronomers to specifically fit their models to its techniques for purely practical reasons. The invention of the differential and integral calculus by Isaac Newton and Gottfried Wilhelm von Leibniz is the most famous example of our first principle. Although Newton presented his

---

[15]The tendency is perhaps reinforced because of our interest in conceptual analysis. To me, one of the most interesting features of philosophy of science is the fact that new methods in various sciences bring to our attention issues and concepts of which we were hitherto unaware. If the philosophy of computational science is a footnote to Plato, it's an exceedingly long one.

[16]See Kline 1972, chap. 22.

results in the *Principia* in traditional geometric form, the subsequent development of classical and celestial mechanics would scarcely have been possible without the differential and integral calculus. The devising of statistical measures of association and inference in the late nineteenth century and early twentieth century by Galton, Pearson, Fisher and others prompted intense interest in agricultural research, genetics, and eugenics; provided epidemiology and biostatistics with their principal tools of trade; and are required areas of study in psychology, sociology, and economics.

Now we have the massive deployment of modeling and simulation on digital computers. This last development is one that is more important than the invention of the calculus in the 1660s, an event that remained unparalleled for almost 300 years, for computer modeling and simulation not only are applicable to a far broader range of applications than are the methods of calculus, but they have introduced a distinctively new, even revolutionary, set of methods into science. The widespread adoption of statistical techniques perhaps comes closest to the global impact of computer modeling, but the latter has the potential to radically reshape large areas of science, such as astrophysics and chemistry, within which statistics play a subsidiary role.

Although traditional mathematical methods will never be completely abandoned, the era of simple, anthropocentric mathematics is over. Standard accounts of how theory is applied to the world have become inappropriate for these new methods because, aside from failing to capture what is distinctive about them, they place constraints on the methods that are misguided. It is our task to explore what should be the replacements for those standard accounts, and we begin with the issue of exactly what it is that should be studied.

3.3 Units of Analysis

Over the years, philosophy of science has taken different kinds of objects as its basic units of analysis. The most common unit is a *theory*. In its most straightforward form, this is considered to be a set of sentences that captures the essential information about some field of inquiry, be it large or small. Examples of such theories abound: probability theory; hydrodynamical theories; status characteristics theories in sociology; classical mechanics in Newton's, Lagrange's, or Hamilton's version; electromagnetic theory; general equilibrium theory in economics; and so on. Axiomatically formulated theories are often the preferred choice of study, and because any consequence of the theory that can be deduced—in principle—from the set of axioms is considered to be implicitly contained in that set, the theory is often identified, indifferently, either with the axioms themselves or with the set of all logical consequences of those axioms. Logically, this closure condition makes for a cleaner unit, although the "in principle" caveat represents a severe idealization. Different from the closure condition is a completeness condition; we want as many as possible—ideally all—of the true claims about the given field to be derivable from the axioms. There are more abstract conceptions of theories, of which the most important is the semantic view, but for now we can think of theories in terms of the sentential view just presented.

Often included as part of a theory, or considered separately, are *scientific laws*, generalizations that apply to and constrain the objects that form the subject matter of a field. Some examples of laws are Maxwell's equations, Newton's three laws, and Snell's law in physics; the Hardy-Weinberg laws in biology; and the law of demand in economics. It is notoriously difficult to provide criteria that distinguish scientific laws from other general statements which appear in theories. In what follows, we do not need to have a solution to that difficulty, for computational science can be highly successful without the need to refer to laws.

Larger units of analysis are entire *research programmes*, imaginatively described by Imre Lakatos.[17] These are sequences of theories based on a core set of inviolate scientific principles which direct research and within which, ideally, each successive theory constitutes progress over its predecessors. Larger still are *paradigms*, famously exploited by Thomas Kuhn,[18] these being collections of core principles, metaphysical assumptions, methodological procedures, and exhibition-quality examples of successful applications.[19] In the other direction, *models* serve an important function in providing representation on a much more restricted scale than theories.[20] Models are usually system-specific representations involving the use of idealizations and approximations, where the choice of the model is frequently guided by heuristic considerations. We shall have more to

[17]See, for example, Lakatos 1970.

[18]As in Kuhn 1970, 1977.

[19]There are even larger units of analysis, bearing long Germanic names, with which we shall not be concerned.

[20]Important early work on models can be found in Suppes 1960, 1962; Wimsatt 1974; Redhead 1980; Achinstein 1968; and Hesse 1966. More recent accounts can be found in Cartwright 1983; and Morgan and Morrison 1999.

say about models, particularly computational models, in sections 3.11 and 3.14.

None of the units just mentioned will exactly serve our purposes. Theories are both too large and too abstract a unit to be useful. They are too large because it is extremely rare that an entire theory is applied to a given task, such as the prediction or description of a system's states. It is selected parts that are used, in a way we shall have to explore in some detail. It is for a similar reason that the logician's conception of theories, which considers them to be simply a deductively closed set of sentences, is too abstract for scientific purposes. Take S as a set of sentences randomly chosen from optics, microeconomics, anthropology, and stereochemistry. As soon as you ask what this is a theory of, the defect is obvious. There must be some coherence to the empirical content of a theory in order for it to count as a scientific theory. In contrast, as we shall see, computational templates can be considered from a purely syntactic perspective.

The study of models suffers from two distinct problems. The first problem is that the term 'model' is used in an enormous variety of ways in science, and to try to give a general characterization of the term is not likely to be rewarding. So in what follows, I shall restrict myself to specific kinds of computational models, with occasional remarks about other kinds. A second problem is that it is difficult to cleanly separate theories from models without imposing artificial restrictions on the former. Various schemata have been proposed for illustrating a hierarchical structure leading from theories through models to data[21] or for having models mediate between theory and data without being derivable from them.[22] Yet mature theories often incorporate models that have been successful in the domain of the theory, and these models are now considered to be a part of the theory.

For example, the theory of probability is no longer identifiable with Kolmogorov's axiomatization, if it ever was. The binomial, multinomial, Poisson, Gaussian, and hypergeometric models, as well as many others, form a central part of probability theory. One might make a case for Kolmogorov's axioms forming the hard core of a research programme, but the theory, and what is essential to it, is far broader than the consequences of those axioms because the theory of stochastic processes forms a central part of contemporary probability theory and this requires significant supplementation of Kolmogorov's original account. Similarly, the theory of

[21]See, e.g., Suppes 1962; Laymon 1982; Bogen and Woodward 1988.
[22]See many of the articles in Morgan and Morrison 1999 for this viewpoint.

classical mechanics should not be identified with Newton's three laws plus his gravitational law, or with various abstract modern versions of them. When you learn classical mechanics, you learn an accretion of general principles, standard models such as the linear harmonic oscillator, and empirical facts about such things as airflow around a moving object producing a frictional force proportional to the velocity, whereas turbulent flow produces one proportional to its square. Because of this intertwining of theory and models, it is preferable to extract other components of the scientific enterprise. I begin with something simple and well known.

3.4 Computational Templates

> In contemporary research almost all the manipulation of the Schrödinger equation is done not analytically but rather by computers using numerical methods... each different potential substituted into the Schrödinger equation typically yields a different problem, requiring a different method of solution.... Moreover... for most physically realistic potentials the Schrödinger equation cannot be solved in analytic form at all... even for the majority of one-dimensional potentials it has become customary to resort to numerical approximation methods, employing a computer.[23]

Take what is arguably the most famous theory of all, Newton's theory of classical mechanics, and one of its central principles, Newton's Second Law. This can be stated in a variety of ways, but in the most elementary case it can be given in the form of a one-dimensional, second-order ordinary differential equation:[24]

$$F = md^2y/dt^2 \qquad\qquad (3.1)$$

The first thing to note about Newton's Second 'Law' is that it is only a schema, what I shall call a *theoretical template*. It describes a very general constraint on the relationship between any force, mass, and acceleration, but to use it in any given case, we need to specify a particular force function, such as a gravitational force, an electrostatic force, a magnetic force, or some other variety of force. If the resulting, more specific,

---

[23]French and Taylor 1998, p. 174.

[24]Although it is possible to represent such equations in more abstract ways—for example, using variable binding operators—these more abstract representations conceal the issues in which we are primarily interested.

equation form is computationally tractable, then we have arrived at a *computational template.*

Theoretical templates are common in mathematically formulated sciences. Probability "theory" is actually a template for which a specific measure, distribution, or density function must be provided before the template can be used. Schrödinger's equation, $\mathbf{H}\Psi = E\Psi$, requires a specification of both the Hamiltonian $\mathbf{H}$ and the state function Ψ in order to be used; various partition functions, such as the grand canonical ensemble partition function $\sum_{i,N} e^{-E_i(N,V)/kT} e^{N\mu/kT}$, must be specified in statistical thermodynamics; with Maxwell's equations $-\nabla \cdot \mathbf{D} = 4\pi\rho$ (Coulomb's law), $\nabla \times \mathbf{H} = (4\pi/c)\mathbf{J} + (1/c)\partial\mathbf{D}/\partial t$ (Ampère's law), $\nabla \times \mathbf{E} + (1/c)\partial\mathbf{B}/\partial t = 0$ (Faraday's law), $\nabla \cdot \mathbf{B} = 0$ (absence of free magnetic poles), where $\mathbf{D}$ is the displacement, $\mathbf{B}$ is the magnetic induction, $\mathbf{H}$ is the magnetic field, $\mathbf{E}$ is the electric field, and $\mathbf{J}$ is the current density—we require a specification of the component physical quantities before they can be brought to bear on a given problem. This despite the fact that four of the examples I have cited are usually given the honorific status of a law of nature.

Computational templates can be found at different levels of abstraction, and more than one step is usually involved in moving from a theoretical template to an application because specific parameter values as well as more general features, such as the forces, must be determined. Thus, if Hooke's law ($F = -ax$) is used as the force function in a particular application describing the restoring force exerted by a stretched spring on a body, then the value of the spring coefficient **a** must be measured or derived from more fundamental principles. Similarly, if a normal distribution is chosen as the appropriate form for some probabilistic application, then values for the mean and the standard deviation must be determined, either by empirical estimation or by derivation from some substantive theory. The most important kind of computational template is found at the first level at which a tractable mathematical form occurs as a result of substitutions into a theoretical template. Above that level, there can be great representational power but no direct application.

We can see how this works for the case of Newton's Second Law. Let us initially take the force function

$$F = GMm/R^2$$

as the gravitational force acting on a body of mass m near the Earth's surface (M is the mass of the Earth and R is its radius). Then the substitution instance of the theoretical template corresponding to Newton's Second Law is

$$GMm/R^2 = md^2y/dt^2, \tag{3.2}$$

and with the initial conditions $y(t_0) = y_0$, $(dy/dt)|_{t_0} = 0$ this computational template is easily solved to give a general solution. Supplemented by the appropriate values for M, m, y_0, and R, a specific solution can be obtained. But the idealizations that underlie this simple mathematical model make it quite unrealistic. These idealizations assume, among other things, that the gravitational force on the body does not vary with its distance from the Earth's surface and that there is no air resistance on the falling body. Suppose we make it a little more realistic by representing the gravitational force as $GMm/(R + y)^2$, where y is the distance of the body from the Earth's surface, and by introducing a velocity-dependent drag force due to air resistance, which we take to be proportional to the square of the body's velocity. We then obtain from the original theoretical template the equation

$$GMm/(R + y)^2 - c\rho s(dy/dt)^2/2 = md^2y/dt^2, \tag{3.3}$$

where ρ is the coefficient of resistance. Now suppose we want to predict the position of this body at a specific time, given zero initial velocity and initial position $y = y_0$. To get that prediction, you have to solve (3.3). But (3.3) has no known analytic solution—the move from (3.2) to (3.3) has converted a second-order, homogeneous, linear ordinary differential equation into one that is second-order, homogeneous, and nonlinear,[25] and the move from linearity to nonlinearity turns syntactically simple tractable mathematics into syntactically simple but analytically intractable mathematics. In order to deal with the problem, we have to provide a numerical solution, usually by converting the differential equation into a difference equation. We thus can see the difference between an analytically solvable computational template and an analytically unsolvable but numerically solvable computational template, both of which are substitution instances of the same theoretical template.

---

[25]The *order* of a differential equation is the order of the highest derivative occurring in the equation. A *linear* differential equation $dy/dx + f(x)y = g(x)$ is one in which all the terms are at most in the first degree in y and its derivatives, and no products of the dependent variables and/or their derivatives occur in the equation. The equation is *homogeneous* if $g(x) = 0$.

This inability to draw out analytically solvable templates from theoretical templates is by no means restricted to physics. It is a common phenomenon because most nonlinear ordinary differential equations and almost all partial differential equations have no known analytic solutions.[26] In population biology, for example, consider the Lotka-Volterra equations (first formulated in 1925).[27] Various equations go under this name; one simple case is

$$dx/dt = ax + bxy$$

$$dy/dt = cy + dxy,$$

where x = population of prey, y = population of predators, a (>0) is the difference between natural birth and death rates for the prey, b (<0) and d (>0) are constants related to chance encounters between prey and predator, and c (<0) gives the natural decline in predators when no prey are available. With initial conditions $x(0) = x_0$, $y(0) = y_0$, there is no known analytic solution to the coupled equation set.

Exactly similar problems arise in quantum mechanics from the use of Schrödinger's equation, where different specifications for the Hamiltonian in the schema $H\Psi = E\Psi$ lead to wide variations in the degree of solvability of the equation. For example, the calculations needed to make quantum mechanical, rather than classical, predictions in chemistry about even very simple reactions, such as the formation of hydrogen molecules when spin and vibration variables are included, are extremely difficult and require massive amounts of computational power. There are no analytic solutions to the time-independent Schrödinger equation applied to the Coulomb potential in three-dimensional, multiple-electron atoms.[28] The integrals involved in the Heisenberg version of the Ising model for ferromagnetism are unsolvable in any explicit way, and although there exist analytic solutions to that model for a one-dimensional lattice, there is no such analytic treatment for the three-dimensional lattice. Moreover, even in the discrete version of that model, for very simple models of lattices with 10^2 nodes in each spatial dimension, there are 2^{10} possible states to sum over in the three-dimensional case. Such examples could be

---

[26]A further source of difficulty, at least in classical mechanics, involves the imposition of nonholomorphic constraints (i.e., constraints on the motion that cannot be represented in the form $f(r_1, \ldots r_n, t) = 0$, where r_i represents the spatial coordinates of the particles comprising the system). For a discussion of these constraints, see Goldstein 1959, pp. 11–14.

[27]This example is taken from Ortega and Poole 1981, chap. 2.

[28]See French and Taylor 1998, p. 219.

multiplied indefinitely, but I hope the point is clear: Clean, abstract presentations of formal schemata disguise the fact that the vast majority of theoretical templates are practically inapplicable in any direct way even to quite simple physical systems.

The inability to draw out computational templates from theoretical templates is by no means restricted to differential equation techniques, and it is the predominance of analytically intractable models that is the primary reason why computational science, which provides a practical means of implementing nonanalytic methods, constitutes a significant and permanent addition to the methods of science. That is why, on the basis of the first of our slogans in section 3.2—that the development of tractable mathematics drives scientific progress—computational science has led to enormous advances in various areas and why, given our second slogan of that section—that scientific models are constrained by the available mathematical methods—computational science suggests that a less abstract approach to scientific method should be used by philosophers, yet one that still provides some good degree of generality, especially across different scientific fields.[29]

In asserting that an equation has no analytic solution we mean that the function which serves as the differand has no known representation in terms of an exact closed form or an infinite series. An important feature of an analytic solution is that these solutions can be written in such a way that when substituted into the original equation, they provide the solution for any values of the variables. In contrast, a numerical solution gives the

---

[29]You might say that this feature of unsolvability is a merely practical matter, and still maintain that as philosophers we should be concerned with what is possible in principle, not what can be done in practice. But investigations into decision problems for differential equations have demonstrated that for any algebraic differential equations (ADE's) (i.e., those of the form

$$P(x, y_1, \cdots, y_m, y_1', \cdots, y_m', \cdots) = 0,$$

where P is a polynomial in all its variables with rational coefficients), it is undecidable whether they have solutions. For example, Jaśkowski 1954 shows there is no algorithm for determining whether a system of ADEs in several dependent variables has a solution in $[0,1]$. Denef and Lipshitz 1984 show that it is undecidable whether there exist analytic solutions for such ADEs in several dependent variables around a local value of x. Further results along these lines, with references, can be found in Denef and Lipshitz 1989. Obviously, we cannot take decidability as a necessary condition for a theory to count as scientifically useful; otherwise we would lose most of our useful fragments of mathematics, but these results do show that there are in principle, as well as practical, restrictions on what we can know to be solvable in physical theories.

solution only at finitely many selected values of the variables. The switch from analytic mathematics to numerical mathematics — and a considerable portion of the computational methods with which we shall be concerned are based on numerical methods — thus has one immediate consequence: We often lose both the great generality and the potentially exact solutions that have traditionally been desired of scientific theories.[30] The first is not a great loss, for applications are ordinarily to specific cases, but the loss of an exact solution introduces issues about the resolving power of the methods that, as we shall see in section 4.3, present novel issues regarding the relation between the templates and the physical computational devices on which they are run. Because in practice we cannot predict with arbitrary accuracy the future states of a system, because data measurements are frequently imprecise, and because all the details of a phenomenon cannot be fully captured even when there exists what we believe to be a complete theory of that phenomenon, the degree to which the resolving power of a computational method fits the available data is a primary concern, even though the inability to arrive at an exact solution is not by itself an objection to numerical methods. The inability to predict future states of a system because of inexactness in the initial conditions is a widely discussed consequence of chaotic systems, but the consequences for computational science are much broader than that.

There is another consequence that follows from our consideration of Newton's Second Law. Systems can be subject to two kinds of influences — general principles and constraints — and the constraints that occur as initial or boundary conditions are as important in using the computational templates as are the 'laws'. In philosophical accounts of science where the laws have been of primary interest, initial and boundary conditions have long been relegated to a firmly secondary status. But in fact, the same equation form or 'law' can be solvable with one kind of constraint yet not with another, and so it is not easy to cleanly separate the 'laws' from the constraints that apply to specific types of applications. Ironically, Laplace[31] endowed his supercalculator with the power to solve any initial value problem governing the time evolution of the state of the universe. But even with those awe-inspiring powers, it does not follow that his calculator would be able to solve the necessary boundary value problems

---

[30]Although numerical solutions are often not exact, this is not an essential feature of them — some numerical solutions give the same exact values as analytic solutions.

[31]In Laplace 1825.

required to be fully successful in showing that the universe is universally deterministic. In the case of templates that are differential equations, it is pointless to consider the equations apart from their related constraints. As one standard source on computational methods puts it: "Even more crucial in determining how to attack a problem numerically is the nature of the problem's boundary conditions.... Usually it is the nature of the boundary conditions that determines which numerical methods will be feasible."[32]

The initial and boundary conditions are usually general rather than particular in form, in the sense that they do not contain any refer- ence to spatiotemporal particulars, but they are not lawlike because, being represented by categorical statements, there are no counterfactuals or subjunctives that flow from the statements of the conditions alone. Nevertheless, initial conditions of a given form can play as great a role as do the underlying laws in determining the broad qualitative features of the phenomena, because different initial conditions lead to qualitatively quite different outcomes even though the same laws are operating. An important example of such conditions involves the motion of a body moving under a central inverse-square law of gravitational attraction. Such a body is constrained by the two relations

$$d^2/d\theta^2(1/r) + 1/r = GM/C^2$$

$$r^2 d\theta/dt = C = r_0 v_0 \sin \omega_0,$$

where G is the universal gravitational constant, M is the mass of the central body, and C is the angular momentum per unit mass. The solution to these equations gives an orbit that is a conic section, the exact form of which is determined by the initial conditions

if $r_0 v_0^2 < 2GM$, the orbit is an ellipse

if $r_0 v_0^2 = 2GM$, the orbit is a parabola

if $r_0 v_0^2 > 2GM$, the orbit is a hyperbola

where r_0 is the initial distance from the central force and v_0 is the initial velocity.

These considerations suggest that although in some contexts, laws are an appropriate focus of interest, they will not be the primary object of investigation in the area of computational science. It is preferable to

---

[32]Press et al. 1988, p. 548.

consider whatever 'laws' there are to be represented by the general equation forms together with precisely specified constraints, and to take as important the computational form of those equations rather than their status as lawlike or not. As a bonus to the present approach, in many of these applications a solution type can be arrived at knowing only the general form of the supplementary factors, be they initial and boundary conditions or parameter values, and our main interest can thus be focused on the first step down from the template. Although there are elements of specificity to computational science, the disunity of scientific method should not be exaggerated.

We have thus arrived at the point where it is apparent that the standard units of philosophical analysis—theories, laws, models, research programs, paradigms—are not the right devices for addressing how much of science is applied to real systems. We have seen how theories and laws fail to fit the bill. But computational templates are not models either, because models are too specific to serve in the roles played by templates. The detailed approximations and idealizations that go into forming models of phenomena make them context-specific and particularistic, in contrast to the generality of templates that gives them their flexibility and degree of independence from subject matter. I shall address the relations between templates and paradigms in section 3.13.

These considerations suggest that although in some contexts and for some purposes the traditional units of analysis in philosophy of science are an appropriate focus of interest, they are not the primary target of investigation in the area of computational science. Computational templates are neither theories nor laws nor models.

3.5 "The Same Equations Have the Same Solutions": Reorganizing the Sciences

> The degree of relationship to real forest fires [of the 'forest fire model'] remains unknown, but the model has been used with some success to model the spreading of measles in the population of the island of Bornholm and of the Faroe Islands. . . . To be consistent with the literature we will use the terminology of trees and fires. The harm in doing so is no greater than in using sandpile language for . . . cellular automata.[33]

[33]Jensen 1998, pp. 65–66.

The drive for computational tractability underlies the enormous importance of a relatively small number of computational templates in the
quantitatively oriented sciences. In physics, there is the mathematically
convenient fact that three fundamental kinds of partial differential
equations—elliptic (e.g., Laplace's equation), parabolic (e.g., the diffusion equation), and hyperbolic (e.g., the wave equation)—are used to
model an enormous variety of physical phenomena. This emphasis on a
few central models is not restricted to physics. The statistician, William
Feller, makes a similar point about statistical models:

> We have here [for the Poisson distribution] a special case of the
> remarkable fact that there exist a few distributions of great uni-
> versality which occur in a surprisingly great variety of problems.
> The three principal distributions, with ramifications throughout
> probability theory, are the binomial distribution, the normal
> distribution . . . and the Poisson distribution.[34]

So, too, in engineering:

> It develops that the equations for most field problems of interest
> to engineers take no more than five or six characteristic forms. It
> therefore appears logical to classify engineering field problems
> according to the form of the characteristic equations and to
> discuss the method of solution of each category as a whole.[35]

Clearly, the practical advantages of this versatility of equation forms is
enormous—science would be vastly more difficult if each distinct phenomenon had a different mathematical representation. As Feynman put it
in his characteristically pithy fashion, "The same equations have the same
solutions."[36] As a master computationalist, Feynman knew that command
of a repertoire of mathematical skills paid off many times over in areas
that were often quite remote from the original applications. From the
philosophical perspective, the ability to use and reuse known equation
forms across disciplinary boundaries is crucial because the emphasis on
sameness of mathematical form has significant consequences for how we
conceive of scientific domains.

Where do the appropriate boundaries lie between the different sciences, and how should we organize them? In various nineteenth-century

---

[34]Feller 1968a, p. 156.

[35]Karplus 1959, p. 11.

[36]Feynman 1965, sec. 12-1. My attention was drawn to this wonderful slogan by Gould
and Tobochnik 1988, p. 620.

treatises, such as those of Auguste Comte and William Whewell,[37] one can find neat diagrams of a scientific hierarchy with physics at the base, chemistry above physics, biology above chemistry, and niches here and there for geology, botany, and other well-established disciplines. Things became a little hazy once the social sciences were reached, but the basic ordering principle was clear: It was the concrete subject matter which determined the boundaries of the specific sciences, and the ordering relation was based on the idea that the fundamental entities of a higher-level science are composed of entities from lower-level sciences. We do not have to agree either with Whewell's and Comte's specific orderings or with the general ordering principle of part/whole inclusion—that if the basic elements of one discipline are made up from entities of another, then the latter is more fundamental to the hierarchy than is the former—to agree that subject matter usually is what counts when we decide how scientific disciplines should be organized.

This focus on the concrete subject matter as determining the boundaries of a science is straightforward if one is an experimentalist. To manipulate the chemical world, you must deal with chemicals, which are often hard and recalcitrant taskmasters, and what can and cannot be done by a given set of techniques is determined by that antecedently fixed subject matter. The same holds in other fields. Try as you might, you cannot affect planetary orbits by musical intervention. Kepler, at least in considering the music of the spheres, just got his subject matters confused.

Things are different in the theoretical realm. Here, reorganization is possible and it happens frequently. The schisms that happen periodically in economics certainly have to do with subject matter, inasmuch as the disputes concern what it is important to study, but they have as much to do with methods and goals. Political economy shears away from econometrics and allies itself with political science because it has goals that are quite peripheral to those working on non-Archimedean measurement structures. Astronomy in Ptolemy's time was inseparable from, in fact was a branch of, mathematics. Subject matter and method are not wholly separable—elementary particle physics and cosmogony now have overlapping theoretical interests due to the common bond of highly energetic interactions (they are both interested in processes that involve extraordinarily high energies; cosmogony in those that happened to occur naturally near the temporal beginnings of our universe, elementary particle

[37]Comte 1869; Whewell 1840.

physics more generally)—but theoretical alliances tend to be flexible. Departmental boundaries can also be fluid for a variety of bureaucratic reasons. Faith and reason once coexisted in departments of religion and philosophy, some of which remain in an uneasy coexistence. The subdisciplines of hydrology, ecology, geology, meteorology and climatology, and some other areas have been merged into departments of environmental science, with somewhat better reason. One can thus find a number of principles to classify the sciences, some of which are congenial to realists, others to anti-realists.

A quite different kind of organization can be based on computational templates. Percolation theory (of which Ising models are a particular example—see section 5.3) can be applied to phenomena as varied as the spread of fungal infections in orchards, the spread of forest fires, the synchronization of firefly flashing, and ferromagnetism. Agent-based models are being applied to systems as varied as financial markets and biological systems developing under evolutionary pressures.[38] Very general models using directed acyclic graphs, g-computational methods, or structural equation models are applied to economic, epidemiological, sociological, and educational phenomena. All of these models transcend the traditional boundaries between the sciences, often quite radically. The contemporary set of methods that goes under the name of "complexity theory" is predicated on the methodological view that a common set of computationally based models is applicable to complex systems in a largely subject-independent manner.

Some phenomena that are covered by the same equations have something in common. For example, the Rayleigh-Taylor instability is found in astrophysics, magnetohydrodynamics, boiling, and detonation. In contrast, there are phenomena that seem to have nothing in common 'physically'. The motion of a flexible string embedded in a thin sheet of rubber is described by the Klein-Gordon equation,

$$\nabla^2 \psi - (1/c^2)\partial^2 \psi / \partial t^2 = k\psi,$$

but so is the wave equation for a relativistic spinless particle in a null electromagnetic field. The statistical models mentioned by Feller cover an astonishing variety of scientific subject matter that in terms of content have nothing in common save for a common abstract structure.[39]

---

[38]Arthur et al. 1997.
[39]For other examples, see Epstein 1997, lecture 4.

This all suggests that, rather than looking to the material subject matter to classify the sciences or to bureaucratic convenience, parts of theoretical science can be reclassified on the basis of which computational templates they use. This kind of reorganization has already happened to a certain extent in some computationally intensive endeavors, including the areas already mentioned, and lies behind much of the movement known generally as complexity theory. Does this mean that the subject matter then becomes irrelevant to the employment of the common templates? Not at all. The primary motivation for using many of these models across disciplinary boundaries is their established computational track record. But as we shall see, the construction and adjustment processes underlying the templates require that attention be paid to the substantive, subject-dependent assumptions that led to their adoption. Without that local, subject-dependent process, the use of these templates is largely unmotivated, and it is this tension between the cross-disciplinary use of the formal templates and their implicit subject dependence that lends them interest as units of analysis.

The idea of an abstract argument form has long been used in logic, although there the orientation has been toward the general form and not the specific instances. A generalization of that idea is Philip Kitcher's concept of a general argument pattern.[40] A general argument pattern, as its name suggests, consists of a skeleton argument form together with sets of filling instructions—sets of directions for filling in the abstract argument form with specific content. An example is the use in post-Daltonian chemistry of atomic weights in explaining why specific compounds of two substances have constant weight ratios of those substances. Computational templates differ from general argument patterns in at least two ways. The issue of mathematical tractability is not a concern for the latter, whereas it is the major focus for the former.[41] Second, general argument patterns provide the basis for Kitcher's account of explanatory unification in an enlightening way. Yet the orientation, judging from the examples, seems to be toward unification within the existing disciplinary organization of science, in contrast to the position advocated here, which is that computational templates provide a clear kind of unity across disciplines. Although unification accounts of explanation are insufficiently realist for my tastes, it would be interesting for advocates of that approach to explore

[40]Introduced in Kitcher 1989.
[41]See the remarks in Kitcher 1989, p. 447, regarding calculations of the chemical bond.

the implications for explanatory unification of the approach suggested here, not the least because it has profound implications for the issue of theoretical reduction. These computational commonalities upset the traditional hierarchy—Gibbs models from physics are used in economics, economic models are used in biology, biological models are used in linguistics, and so on. "Theory" longer parallels subject matter, although it has not been cast entirely adrift.

3.6 Template Construction

> Many theories of dispersion have been proposed. The first were imperfect, and contained but little truth. Then came that of Helmholtz, and this in its turn was modified in different ways. . . . But the remarkable thing is that all the scientists who followed Helmholtz obtain the same equation, although their starting points were to all appearances widely separated.[42]

Thus far, we have considered basic equation forms such as Newton's Second Law as fundamental. However, most well-established theories construct their templates from a combination of basic principles, idealizations, approximations, and analogical reasoning. Perhaps the attitude that templates are to be treated as primitive comes from the emphasis on axiomatic formulations of theories, within which the focus is on the basic principles of the theory. But whatever the reason, examining the construction process allows us to highlight five features of templates that are crucial when they need to be revised. The first is that identical templates can be derived on the basis of completely different and sometimes incompatible assumptions. The second is that even bare equation forms, treated as syntactic objects upon which computations are to be performed, come with an intended interpretation. The third feature is that the construction process provides an initial justification which is built into the template, and although this justification can be overridden by contrary empirical evidence, it requires a specific counterargument against particular elements of the construction process. The fourth feature is that subject-specific knowledge is required not only in the construction of these models but also in their evaluation and adjustment. The fifth feature is that we usually know in advance how the template needs to be revised when it does not fit the data.

[42]Poincaré 1952, p. 162.

Once these five facts have been highlighted, it is much easier to see how a defective template can be revised, rather than leaving the adjustments to a mysterious process of scientific intuition, to convention, or to "pragmatics." I shall illustrate the construction process by appeal to three separate constructions of the same general template. Although this will of necessity be case specific, the points I want to make are transferable to almost any standard equation form used in scientific modeling.

The diffusion or heat equation is widely used in modeling physical phenomena, and solution methods for specified boundary conditions are a stock-in-trade of applied mathematics. For application to specific problems, the use of the diffusion equation must be justified, and a common way of doing this is to construct the differential equation from a set of more basic assumptions.

The First Derivation: The Heat-as-a-Fluid Analogy

Consider an insulated conductor of heat in the form of a cylindrical rod with cross-sectional area a and temperature function $u(x, t)$. Then

(1) Quantity of heat in section $= sa\Delta x\rho u$ where s = specific heat, ρ = density.

Hence

(2) Temporal flow of heat in section $= sa\Delta x\rho \partial u/\partial t$.

Also,

(3) Flow of heat across area a along rod $= -ka\partial u/\partial x$.

Hence

(4) Net flow of heat in section $\Delta x = -(ka\partial u/\partial x)|_{x=x} - (-ka\partial u/\partial x)|_{x=x+\Delta x}$.

Thus

(5) $sa\Delta x\rho \partial u/\partial t = ka(\partial u/\partial x|_{x=x+\Delta x} - \partial u/\partial x|_{x=x}) = ka\Delta(\partial u/\partial x)$.

(6) So as $\Delta x \to 0$,

$\partial u/\partial t = k/s\rho \, \partial^2 u/\partial x^2$.

The components in this derivation are easy to identify. Basic physical principles have been used explicitly in (1) and (3), and implicitly in (4), where a conservation principle is applied. This application of the conservation principle is possible because of the idealization of a perfectly insulated conductor, which justifies the unstated boundary condition $\partial u / \partial r = 0$ at the surface of the conductor, where r is any radial direction orthogonal to the surface. Further physical idealizations are (a) a perfectly cylindrical conductor (more exactly, one with a uniform cross section), and (b) constant and uniform specific heat, density, and k. (If k is a function of temperature, for example, then (4) leads to a nonlinear version of the diffusion equation.) A mathematical/physical idealization occurs in (6), to ensure that u is sufficiently smooth to allow the limit process to be justified. Equation (2) merely uses a definition of flow, whereas (5) is an identity. Finally, an analogical assumption is used at the outset, to the effect that heat can be considered as an entity that flows.

The Second Derivation: Molecular Motion as a Random Walk

Here we have a model with molecules moving in discrete steps from one lattice node to another.

Let $P_n(i) = $ probability that molecule is at location i after n steps.

Then $P_n(i) = P(\text{left})P_{n-1}(i+1) + P(\text{right})P_{n-1}(i-1)$
$$= q P_{n-1}(i+1) + p P_{n-1}(i-1).$$

Let $\tau = $ time between collisions and $a = $ lattice spacing.

Then time $t = n\tau$, position $x = ia$, $P_n(i) = aP(x, t)$, where $P(x, t)$ is the probability density.

We thus have $P(x, t) = qP(x + a, t - \tau) + pP(x - a, t - \tau)$

If $p = q = \frac{1}{2}$, then

$$\tau^{-1}[P(x, t) - P(x, t - \tau)] = (a^2/2\tau a^2)[P(x + a, t - \tau) \\ -2P(x, t - \tau) + P(x - a, t - \tau)]. \tag{1}$$

Using the Taylor series expansion

$$P(x, t - \tau) = P(x_0, t - \tau) + (\partial P(x_0)/\partial x)(x - x_0) \\ + \tfrac{1}{2}(\partial^2 P(x_0)/\partial x^2)(x - x_0)^2 + O(x_0^3)$$

successively for $x_0 = x + a$, $x_0 = x - a$, and substituting in the right-hand side of (1), we have

$$\tau^{-1}[P(x, t) - P(x, t - \tau)] = \\ (a^2/2\tau)a^{-2}[(\partial P(x + a, t - \tau)/\partial x - \partial P(x - a, t - \tau)/\partial x)a \\ + a^2/2(\partial^2 P(x + a, t - \tau)/\partial x^2 + \partial^2 P(x - a, t - \tau)/\partial x^2) + O(a^3)].$$

As $a \to 0$, $\tau \to 0$ with $a^2/2\tau = K$ finite, we have

$$\partial P(x,t)/\partial t = K\partial^2 P(x,t)/\partial x^2,$$

the diffusion equation for P.

The components of this derivation are equally easy to state. Physical principles are used implicitly only in the immediate consequences of adopting the molecular collision model underlying the derivation. An explicit physical constraint involving the dynamic equilibrium of the gas molecules is used to justify the equality of the probabilities for moves of the molecules to the right and left. The rest of the derivation involves mathematical identities, order-of-magnitude assumptions, and the mathematical/physical assumption that $a^2/2\tau$ remains finite in the limit process. A significant, and clearly false, physical idealization is that the length of the path between collisions, a, is a constant, as is the velocity of the molecule (which latter assumption uses the conservation of momentum and idealized perfectly elastic collisions). In addition, we assume that successive collisions are statistically independent.

The Third Derivation: Fluid Flow as a General Concept

This derivation proceeds from two general assumptions:

(1) That it is possible to consider the quantity represented by the dependent variable as a compressible 'fluid'.
(2) That it makes no physical sense to speak of forces on an element of the 'fluid'.

Using (2), the flow is not caused by external forces or pressure gradients. Instead, it is due to a density gradient. Thus, if J is the flow per centimeter at each point, then

(3) $J = -k$ grad P,

where k is the diffusion constant, P is the quantity of heat per unit volume, and

grad $P = \nabla P = (\partial P/\partial x)i + (\partial P/\partial y)j + (\partial P/\partial z)k.$

Next, invoke a conservation principle, the equation of continuity:

(4) $\partial P/\partial t = -$div $J = -\nabla \cdot J = \partial J x/\partial x + \partial J y/\partial y + \partial J z/\partial z.$

(If fluid were lost, this would not hold.)

Thus, from (3) and (4)

$$\partial P/\partial t = k\nabla^2 P.$$

This highly abstract derivation invokes a fundamental physical principle at (4). Idealizations similar to those in the first derivation are used, except that no assumptions of smoothness are required, because the derivation uses no (explicit) discrete approximations. The use of analogical reasoning is evident in (1), with a consequent and subsidiary use in (2). However (3) requires some physical features of the 'fluid'—specifically, that it has a density (gradient). In contrast, (4) is perfectly general, and holds independently of the type of fluid, as long as the idealizations hold also.

We see in each of these derivations that only highly selective parts of the 'theory of heat' are used in the construction process. Moreover, the other assumptions are made specifically for the model at hand and are not drawn from a general theory, even though in some cases they can be transferred to other types of systems.

3.7 Correction Sets, Interpretation, and Justification

At some point a computational template has to be tested against empirical data. What happens when it fails to fit those data? How do we adjust the template? The answers to these questions are closely connected with a standard distinction that for many years reigned in the philosophy of science, the distinction between the context of discovery of a scientific theory or hypothesis and the context of justification of that theory or hypothesis. The context of discovery was held to be one of little relevance to the testing of theories. It was largely of psychological interest, a realm of conjectures made on the basis of scientific intuition and occasional genius. The context of justification constituted the area of real philosophical interest, for it was there, by comparison, that theories were objectively tested and either confirmed or rejected. This distinction came under attack, primarily from sociologically oriented writers, for reasons that would take us too far afield to address here. For our purposes, there is another twofold classification that better represents the use of computational templates and models and that can provide us with the basis for answers to our questions. We can call the two aspects the *process of construction* and the *process of adjustment*. The processes are intimately related.

When the predictions from a template and the data are at odds with one another, the Duhem-Quine thesis[43] notes that because the template is embedded in a network of auxiliary assumptions, any of those assumptions could, in principle, be blamed for the failure. Logically, this thesis is completely trivial.[44] Whatever philosophical interest it has must lie in the procedures that are suggested for revising the template or the auxiliary assumptions. A standard suggestion is one of epistemological holism, the view that the justification of any given assumption cannot be separated from the justification of any other. Whatever its merits may be for theories conceived of as sets of sentences, epistemic holism misdescribes the process of template correction. The mistake arises from viewing the template from what we may call the *no-ownership perspective*. Within that perspective, the template is conceived of as a found item, and if it is subsequently discovered to be false, the task of repairing it is begun with no clues available about why it has the form that it does and no guidance from the template itself as to how it should be adjusted. This picture is a significant misrepresentation of how templates are tested and refined because the constraints on adjustments are far tighter than the holist position suggests. We usually expect, in advance of testing the template, that the data will deviate from the template's predictions and we know which of the assumptions that went into the construction of the template are less than adequate for the particular application at hand. Because of that expectation we often have, before the testing takes place, a plan for improving the template.

The existence of this plan means that the temporal scheme

hypothesis $\Rightarrow$ conflict with data $\Rightarrow$ decision concerning adjustment

is in many cases an inaccurate description of the adjustment process. No doubt there are situations in which a hypothesis is thrown to the data with full confidence that it is adequate, only for surprise to set in when it turns out to be faulty. Most mathematical modelers are a little more wordly-wise about their models than this and know from the outset that their models are defective in certain aspects. More important is the fact that the improvements which result from correcting the template are frequently not the result of conventional or pragmatic decisions but, as we shall see in section 3.11, are made on the basis of substantive, subject-matter-dependent content.

---

[43]See, e.g., Duhem 1906; Quine 1951.
[44]There are other, much less trivial, forms of holism that I do not address here.

A more accurate account of template correction goes along the following lines. The assumptions used in the template construction can be represented by a quintuple <Ontology, Idealizations, Abstractions, Constraints (which include laws), Approximations>, which I shall call the *construction assumptions*. The ontology specifies the kinds of objects represented in the template and the nature of the properties and relations occurring in it, often including mechanisms that operate between and within systems.[45] Unlike the other four components of the construction assumptions, which can be purely formal, the ontology is necessary in order to turn a general model into a model of a specific system type and to provide a basis for the other construction assumptions. Although a multinomial model in statistics can be a model of many different phenomena, it is a model of *dice tossing* only when the objects are deliberately specified as dice. The model would be untestable if the ontology was not specified, because we would not know if real dice had the properties imputed to them by the model.

The second main component of a template that is often formulated during the construction process, but is not used initially, is its *correction set*. The principal purpose of the correction set is to indicate, in advance, the ways in which the computational template will need to be adjusted when (not if) it fails to match the empirical data. The correction set is not used in the initial construction for a variety of reasons. It might be because data are needed to bring out how far the template's predictions are from the system's behavior. It might be because the mathematical techniques for bringing the correction set into play have not yet been developed. Or it might be because the computing power required for using a more refined template is not yet available. Unlike the construction assumptions, for which solvability is the primary initial constraint, increased accuracy under the constraint of solvability is the correction set's aim. The components of the correction set will vary depending upon the construction assumptions that were used for the computational template, but the most important constituents are

1. Relaxation of idealizations (e.g., moving from considering the Sun as a perfect sphere to considering it as an oblate spheroid)
2. Relaxation of abstractions (e.g., adding previously omitted variables to the model, such as introducing a treatment of friction into

[45]A valuable collection of articles on social mechanisms is Hedström and Swedberg 1998.

a template that initially represented motion only over a frictionless surface; increasing the number of individuals in an ecological simulation; or adding more stars to a galactic simulation)

3. Relaxation of constraints (e.g., moving from a conductor within which heat is conserved to one that has heat flowing across the boundary)

4. Refinement of approximations (e.g., reducing the spatial grid size in finite difference approximations to continuous models or replacing an aggregated variable with precise values)

5. Changes in the ontology of the system. This is a radical move but often occurs when the computational template is applied to a different subject matter (e.g., when we apply a contagion model of an epidemic to the spread of revolutionary ideas in a poor urban society). Changes in the ontology will often require changes in the other elements of the correction set and come very close to starting a new construction process.

The original ontology often plays an important role in formulating components of the correction set. Here is one example. Some years ago an experiment was conducted consisting of 4.38 million tosses of various dice—by hand! It was discovered that there were measurable differences between the results obtained from dime store dice and those obtained from professional quality dice. In an appendix to the published paper, a model is given for the dime-store dice and a linear correction for the physical effects of the holes drilled for the pips is provided.[46] In contrast, the Las Vegas dice use pip filler whose density matches that of the die's body, and this leaves unaltered the symmetry possessed by the blank die. Without that underlying physical model of the dice, it would not be possible to explain and correct for the variations from the model in the performance of the different dice. In the theoretical realm, the ontology will often place restrictions on which mathematical operations are permissible within the model when the model is interpreted realistically.

Introducing the correction set allows a better representation of template adjustment along these lines:

construction of template ⇒ assessment of assumptions ⇒ formulation of correction set ⇒ comparison with data ⇒ adjustment of template according to the correction set.

---

[46]Iverson et al. 1971.

In extreme cases, the process can lead to abandoning the template altogether, but this is unusual, especially where the template has already proved its worth.

The view that the process of adjusting defective models is driven primarily by pragmatic considerations may seem plausible when the models are viewed from a purely formal perspective. Yet the idea that we can separate the formalism from its interpretation can be seen to be incorrect once we replace the context-of-discovery standpoint with the process-of-construction perspective. We can call the view that the template is a mere piece of formalism upon which some interpretation can be imposed, together with the idea that the interpretation can be removed and replaced with some other interpretation, the *detachable interpretation* view. To adopt the detachable interpretation view is to deeply misrepresent the epistemic status that templates have in scientific work. If we take the basic template forms as primitive bits of syntax, then we omit the fact that these bits of syntax come, via the construction of those templates, with both an intended interpretation and an associated justification embedded in them. To peel off the intended interpretation is to remove the justification we have for adopting the equations. For example, in the first derivation of the diffusion equation we examined, the interpretation of the function $u(x, t)$ represented the temperature gradient in a perfectly insulated cylindrical conductor, and this interpretation is central to the decision whether or not the diffusion equation is a correct representation of the flow of heat in a given metal bar. If we want to use the diffusion equation to represent neutron diffusion in a material, the interpretation of $u(x, t)$ just given is of no use in justifying the adoption of the diffusion template for that use. Once we have removed the heat interpretation, the original justification goes with it. Reinterpreting the diffusion template in terms of neutron movements does preserve truth, but it leaves the template as a floating conjecture, devoid of independent justification.

The detachable interpretation position is often presented in terms of the idea of a theory having an 'intended interpretation' being indefensible. That view gains traction from the no-ownership perspective, a perspective that misrepresents how templates are used. When a different interpretation is put by us onto a template, the applications of that template will ordinarily be different from those associated with the original, intended, interpretation. It must then be shown by the new owner that the second interpretation provides a justification for the construction and application of the template in the new circumstances, not just that the reinterpreted equation would be true just in case the original was.

The templates are thus not mere conjectures but objects for which a separate justification for each idealization, approximation, and physical principle is often available, and those justifications transfer to the use of the template. Moreover, when there is a lack of fit between the predictions of the template and the data, rather than simply abandoning the former, explicit attention to the correction set is frequently the best guide to adjusting and retaining the template. In arriving at solvable equations, especially when using numerical methods, a variety of further approximations and idealizations will need to be introduced. Frequently, the solutions from that initial template will be inconsistent with empirical data, and if adjustments are required in the equation forms (rather than in the further approximations and idealizations), these adjustments are usually neither ad hoc nor guided by pragmatic considerations such as simplicity, but are the result of improvements in inadequate initial approximations and idealizations. It is an important fact that such inadequacies are known in advance, and that models known to be false are tested with an eye not to falsifying and rejecting them but to improving them.

Of course the hypothetico-deductivist can reply that template building is just a deductive process from very basic principles, and that what we have pointed out is simply that the equations we take as the starting point are amenable to a deductive construction. This is certainly true, but to leave it at this would be to miss the point. Epistemological holists frequently, perhaps universally, write as if incorrect predictions leave everything, logically, open to revision, with only pragmatic considerations motivating the decision. This is, for most situations, a serious under-description of the situation. Template builders usually have a quite refined sense of which of the components that go into the construction of the template are well justified and which are not. They often have, in advance of testing, a clear idea of which parts of the construction will be the first to be revised when the template fails to give accurate predictions. Parts of the construction process are selectively targeted for revision, and this is done on the basis of knowledge specific to the template and to the application. We have already seen this with the different construction processes that lead to the diffusion equation template. Because a particular construction process will be chosen on the basis of the underlying ontology of the system, the revisions will take a different form for different construction processes even though the equations involved are formally the same.

Perhaps the attitude that equation forms are to be treated as primitive comes from the emphasis on axiomatic formulations of theories, within which the testing procedure is aimed primarily at the axioms of the theory. But whatever the reason, focusing on the construction process allows us to highlight two features of scientific models that are crucial as guides to revising incorrect theories. The first is that even bare equation forms, treated as syntactic objects, come with an intended interpretation. The second feature is that the construction process itself provides an initial justification for the theory that is built in and that, although it can be overridden by contrary empirical evidence, requires a specific counterargument against particular elements of the construction process. Once these two facts have been highlighted, it is much easier to see how systematic methods can be applied to revise a defective template, rather than leaving the adjustments to a mysterious process of scientific intuition.

There are echoes of Lakatos's famous advocacy of hard cores of research programmes here.[47] There is, however, a distinct difference in perspective between our position and his. Aside from the finer-grained level of analysis—the process is internal to specific computational templates rather than to entire research programmes—the constraints from solvability are not much emphasized in Lakatos's views on scientific research programs, whereas they are an important constraint on the templates here considered.[48] This is one reason why a purely realist interpretation of these models is often implausible—truth is not the only constraint; solvability is as well. But to emphasize the importance of solvability is not to adopt instrumentalism. Overall, the model is almost always false, which is to say that the predictions drawn from it rarely are exactly true. And by asserting its overall falsity, we have already rejected an instrumentalist interpretation. Yet despite this overall falsity, parts of the model often are true and some of the terms do refer. This is why one needs to be a selective realist, a position to which we now turn.

3.8 Selective Realism

"It's not fake anything, it's real plastic."—Andy Warhol

It is hard to find common ground upon which realists and anti-realists agree, and because of this, disputes about scientific realism tend to result in

---

[47]Lakatos 1970.
[48]There are some remarks by Lakatos to this effect—see footnote 17.

a frustrating standoff or accusations of question-begging. This is especially noticeable when the dispute is over whether unobservable entities exist. By drawing the line between what is observable and what is unobservable at some self-interested place, a conservative realist can seem indistinguishable from a liberal empiricist, or a tightfisted empiricist can simply refuse to accept evidence that, to a realist, appears highly persuasive. Advocates of realism often desire an all-purpose criterion of reality, applicable across all subject matters. An example of this is the view, suitably hedged, that any empirically successful theory must correctly represent the world. This is extravagant, for realists must acknowledge that there are theories whose success is not due to their truth and, more commonly, that successful theories have parts which we do not want to interpret realistically. The quantum realm abounds with examples of the former; for the latter, Newtonian gravitational theory famously induced nervousness about the reality of the gravitational force, even from Newton himself. In contrast, almost everyone was a realist about distances (although in different ways, depending upon whether one was committed to absolute or relational accounts of spatial relations) and most, although not all, were willing to be realists about mass.

Realism is often taken to be a position about theories taken in abstraction from our attitude toward them. It is said to be about whether those theories are true or false, or whether certain terms of well-confirmed theories refer to real entities. This is what we called in the last section the no-ownership perspective. But realists are rarely so unselective about, or so distanced from, the contents of their theories. Realism is ordinarily an attitude held by an individual or by a group toward parts of a theory, and the realist commitments are not usually read straight off the theory itself. A particularly influential account which claims, in contrast, that they can be so read from the theory is Quine's famous theory of ontological commitment: "To be is to be the value of a variable" is the slogan; a more refined statement is "In general, entities of a given sort are assumed by a theory if and only if some of them must be counted among the values of the variables in order that the statements affirmed in the theory be true."[49] Yet even this doctrine does not commit us to having our theories dictate our ontology to us. For Quine goes on to say:

> One way in which a man may fail to share the ontological commitments of his discourse is, obviously, by taking an attitude

---

[49]Quine 1961a, p.103.

of frivolity. The parent who tells the Cinderella story is no more committed to admitting a fairy godmother and a pumpkin coach into his own ontology than to admitting the story as true. Another and more serious case in which a man frees himself from ontological commitments of his discourse is this: he shows how some particular use which he makes of quantification, involving a prima facie commitment to certain objects, can be expanded into an idiom innocent of such commitments.[50]

An admirably wholesome concession to our freedom to choose our ontological associates, and by no means the only way to specify which parts of one's templates are to be construed realistically.

Let us call *selective entity realism* an attitude toward a template held by some individual within which certain specified terms of the template are asserted by that individual to correspond to parts of a specified system. This allows that an individual's selective entity realism about a given entity may or may not be justified or rational. It is possible to deny that any of the terms are to be construed realistically. In that case the template will be part of what is colloquially called a "math game" within which the assumptions behind the template are so far removed from reality that construing it as applying to the world would be foolish. Math games can be an important part of the development of computational methods, for they allow theoretical progress to occur before we understand how to relax the assumptions behind the template to a point where they become realistic.

In addition to his discriminating attitude toward the basic terms of a template, a selective realist will have to reject certain closure conditions that are imposed on the formalism for mathematical convenience. A standard, logically inspired, view of theories identifies a theory, as we saw in section 3.3, with the set of all deductive consequences of its axioms. That sort of closure is acceptable for purposes of truth preservation (if you think the axioms are all true, you have to accept that any consequence drawn from them is true, too), but in order to preserve that kind of closure while not letting our realist commitments run rampant, certain other closure conditions that are often included as part of the background mathematical apparatus have to be rejected. For example, it is standardly assumed in most theories of mathematical physics that all the derivatives of a function exist. But arbitrary derivatives of a function, such as the

---

[50]Quine 1961a, p. 103.

eighth derivative, are usually physically meaningless. Most equations of physical interest have an order no higher than second. (One exception is the displacement of the end of an elastic bar rigidly fixed at the other end, which is described by a fourth-order ordinary differential equation.)[51] The same thing occurs with ordinary arithmetical operations. What physical referent does the square root of charge have, for example? We have here a tension between the need for mathematical simplicity and the commitment to physical content of the theory.[52] The obvious solution is to be explicit about which parts of the theory are to be construed realistically and which merely instrumentally. The theory itself is mute—we must speak on its behalf.

3.9 Further Consequences

> I am tired of hearing neo-instrumentalists and neo-realists say that scientific realism is false because all of the theories and theoretical tools we use in science are false or untrustworthy. I am tired of hearing representatives of the new wave in sociology of science say that all of the products of science, from theories to data, are social constructs, so we should all be social relativists rather than scientific realists. I am tired of hearing some philosophers who call themselves "scientific realists" talk as if their scientific realism has any close connection with those views held by scientists who would call themselves realists,...It may be revealing to see how much we can learn from tools which are known to be defective.[53]

These refreshing—if uncharacteristically exasperated—comments of Bill Wimsatt draw our attention to an important fact: that users are often clear-eyed about the fact that their models are false. Wimsatt's article, which should be read be all who are interested in modeling, deals with that fact in a particular way. Here I want to use Wimsatt's point to illuminate something slightly different, but still in the spirit of his remarks.

---

[51]Not much weight should be put on these distinctions, however, because any *n*th-order ODE can be reduced to a system of first-order ODEs.

[52]Hilbert 1926 has illuminating things to say about the move to mathematical closure and its consequences for formalism.

[53]Bill Wimsatt, unpublished draft of "False Models as a Means to Truer Theories," University of Chicago, October 17, 1984. Reproduced with Wimsatt's permission. The passage was omitted from the published version, Wimsatt 1987.

A number of different reasons support the position that false theories or models can accurately capture observed phenomena. I shall concentrate on only one, albeit an important, way in which this can happen. This is when an ontology is used within a template that is quite different from the ontology underlying the real system.

The well-known underdetermination arguments that we shall discuss further in section 5.1 present it as a simple fact that there can be radically different ontologies for the same formal model. In contrast, the construction processes laid out for the diffusion equation in section 3.6 explain how this can happen. The different processes of abstraction, idealization, approximation, and simplification that are applied to different ontological starting points show how exactly the same template can be derived from those fundamentally different bases. Because it is the template that serves as the basis for the predictions which are compared with observations, it is immediately evident that major differences in the starting points can lead to the same predictively successful template.[54] Once we understand how ontologically deviant templates can work so well, we can distinguish between different kinds of false models.

In some cases an explicit disclaimer is made from the outset that the template is not to be interpreted realistically and the falsity is so blatant that there is no intention of using a correction set. One example of such radically false but successful templates involves using classical mechanical representations of the movements of atoms in molecules. This accounts well for molecular chirality, ring conformation and steric strain, and the common use of the Born-Oppenheim approximation even though it involves the explicit adoption of a model that all users recognize as false. In other cases, the position of selective realism allows the user to be explicit about which parts of the false template he holds to be false and which he holds to be true. Unless we draw a sharp distinction on the basis of users' intentions, based on selective realism between the first, blatantly false, type of template and those where some effort will be made to correct deficiencies through use of the correction set, it is impossible to distinguish the important differences between these two kinds of false representation.

A second conclusion we can draw is that because the idealizations and approximations used in the construction assumptions enter directly

---

[54]Of course, this is by no means the only reason why predictive success can occur with false models. For example, mistaking correlations induced by a common cause for a direct causal relation is another source of successful but false models.

into the construction of the template, no 'ceteris paribus' conditions are required for the statement of the equation, because the restrictions on the use of the law have already been explicitly incorporated into the justificatory conditions for the equation. The only feature of the construction process underlying the diffusion equation that resembles the use of ceteris paribus conditions is the requirement in the first and third derivations that the quantity under consideration can be considered as a fluid. This assumption is needed to allow the use of definitions of flow. One characteristic of a fluid is that a constant rate of shear stress produces a constant shear force, and this can be used to distinguish fluids from elastic solids, for example, in a way that does not use any of the fundamental physical principles employed in the construction of the diffusion equation. Such an assumption relies on an analogy and the 'ceteris paribus' condition here is simply the assumption that the analogy is valid.

As I have noted elsewhere,[55] the term 'ceteris paribus' is used in at least four different ways in theories that have causal content. There is a fifth, noncausal, use in the context of template employment, where 'ceteris paribus' is simply the claim that the construction assumptions underlying the computational template are true. It has been said that a characteristic feature of ceteris paribus conditions is that they cannot be stated finitistically or, alternatively but differently, that they cannot be given a precise statement. I am not in a position to answer that point in general. I am sure that there are such uses. But in the case of templates that have been explicitly constructed, there is no vagueness about the conditions under which they hold. If the construction is correct, then the assumptions which enter into the construction process are sufficient for that template to be accurate. Often, the assumptions are not necessary—a fact that is evident from the multiple derivations. But the construction process is usually finitistic and explicit, and hence the ceteris paribus claim, construed in our fifth sense, is finitely statable and precise. To suggest otherwise is to introduce unwarranted confusions.

So the principal conclusion to be drawn from these considerations is that productive uses of the hypothetico-deductive method should, when possible, take into account the origins of the templates that are under test, and that the justification of the template is intertwined with its construction.

---

[55]Humphreys 1989.

3.10 Templates Are Not Always Built on Laws or Theories

The templates we have considered so far have been constructed on the basis of assumptions, some of which are justified by an explicit appeal to scientific laws. In the first and third derivations, each contained a law, the conservation of heat and the conservation of fluid, respectively. Other templates make no use of laws. Consider an utterly simple construction, for which the resulting computational template is that of a Poisson process.[56] Suppose we are willing to make the following assumptions about the stochastic process:

1. During a small interval of time, the probability of observing one event in that interval is proportional to the length of that interval.
2. The probability of two or more events in a small interval is small, and goes to zero rapidly as the length of the interval goes to zero.
3. The number of events in a given time interval is independent of the number of events in any disjoint interval.
4. The number of events in a given interval depends only upon the length of the interval and not its location.

These are informal assumptions, but ones that can be given precise mathematical representations. From those representations it is possible to derive the exact form of the probability distribution covering the output from the process: $P(N = k) = e^{-\lambda}\lambda^k/k!$. This template is of great generality, and it has been used to represent phenomena as varied as the number of organisms per unit volume in a dilute solution, the number of telephone calls passing through a switchboard, the number of cars arriving per unit time at a remote border post, the number of radioactive disintegrations per unit time, chromosome interchanges in cells, flying bomb hits on London during World War II, the number of fish caught per hour on a given river, the rate of lightbulb failures, and many others.[57] Some of these applications, such as the flying bomb hits, provide reasonably accurate predictions of the observed data, whereas others give at best approximations to the empirical results. Leaving aside the issue of accuracy, one conclusion to be drawn from this example is that templates can be constructed in the absence of anything that looks like a systematic theory about a specific domain of objects. It is implausible to claim that in the

---

[56]For a simple exposition of Poisson processes, see any standard text on probability. One clear account can be found in Barr and Zehna 1971, pp. 344ff.

[57]See, e.g., Huntsberger and Billingsley 1973, p. 122; Feller 1968a, pp. 156ff.

case of the flying bomb application there exists a systematic theory of aerodynamics that justifies the use of each assumption, and in this and in other cases, the Poisson assumptions can be justified purely on the basis of empirical counts. Theories need play no role in constructing templates.

And so also in the case of laws. In many cases of template construction, laws will enter into the construction process, but in the Poisson example just described it is not plausible in the least that any of the assumptions given count as laws. Although one might say that the individual outcomes are explained by virtue of falling under a lawlike distribution, if the fact to be explained is that the system itself follows a Poisson distribution, then that fact is not understood in virtue of its being the conclusion in a derivation within which at least one law plays an essential role. Furthermore, even if we consider the distribution that covers the process itself to be a law, that law is not itself understood in terms of further laws, but in terms of structural facts about the system. This clearly indicates that scientific understanding can take place in the absence of laws.[58]

The point is not that there are no scientific laws—there are.[59] There are fundamental laws such as the conservation of charge in elementary particle reactions and decays, the conservation of isospin in strong interactions, and the conservation of charge conjugation parity in strong and electromagnetic interactions, to name only three. These laws are, to the best of our current knowledge, true and hence exceptionless. There are others, such as the conservation of baryon number, which hold universally except for situations that can be explicitly specified—in this case proton decay—and these exceptions have not to date been observed. Combine this with our point that accurate and widely used templates can be constructed without the need for laws, and the combination of the two leads us to the conclusion that there are laws but that they need not be the starting point for applying science to the world.

We can now address a central issue for templates. How is it that exactly the same template can be successfully used on completely different subject matters? This is a different question than the one that famously asks "What explains the unreasonable effectiveness of mathematics in representing the world?" That older question cannot simply be

---

[58]For more on this, see Humphreys 2000.

[59]With, perhaps, the additional twist that, as van Fraassen (1989) has suggested, talk of laws in fundamental physics should be replaced by talk of symmetries.

about why mathematics applies to the world, for there would be a straightforward answer to it, which is that for the most part, it does not. Mathematical models in science generally do not apply directly to the world, but only to a refined version of it. Because the space of consistent mathematical structures is much larger than the space of idealized physical structures, it is not at all surprising that one can find, or construct, objects in the former realm that apply to the latter. Accept the idealizations, approximations, and abstractions involved in the physical models, and it is inevitable that some mathematical model will work.

A slightly more interesting interpretation of what is meant by the question suggests that what is surprising is that parts of the world are such that relatively simple mathematical models fit them closely, even without much idealization. An answer to that version of the question, perhaps not much more interesting than the first answer, is that a selection bias has been in operation and that the era of simple mathematics effectively modeling parts of the world is drawing to a close. It is possible that new areas of investigation will lend themselves to simple models, but the evidence is that within existing areas of investigation, the domain of simple models has been extensively mined to the point where the rewards are slim.

There is a third version of the older question: Why do parts of mathematics that were invented for the purposes of pure mathematics model the natural world so well? One answer to this question is simply a variant of our answer to the first question: Much of mathematics has been constructed, and it is perhaps not surprising that we can find some parts of it that happen to apply to selected features of the world. A related and more accurate answer relies on a variant of our second slogan from section 3.2 — we simply force-fit our models to the available mathematics because which forms of mathematics are available constitutes one of the constraints within which we must work.

My question is different from these older questions. Given that mathematics does apply to the world, why do the same mathematical models apply to parts of the world that seem in many respects to be completely different from one another? On a hypothetico-deductive approach to templates, this fact appears quite mysterious. In contrast, on the constructivist account of templates we have suggested there is a ready explanation. Templates are usually not constructed on the basis of specific theories, but by using components that are highly general, such as conservation laws and mathematical approximations, and it would be a mistake to identify many of these components with a specific theory or subject matter. The Poisson derivation is an extreme example of this, for it uses only purely structural features of the systems, and those features can often be justified on the basis of simple empirical data without any need for theory—for example, the fact that the probability of two cars arriving at a remote border post within a short time interval is small.[60] That is why such templates are applicable across so many different domains. The role of subject-specific theory and laws is nonexistent in these extreme cases, except when such things are used to justify the adoption of one or more of the assumptions used in the construction. Most template constructions are more specific than this, but in the case of the diffusion equation, the independence from theory is most clearly seen in the third derivation, where nothing that is theory-specific is used. Even in the more specific constructions, it is the fact that mere fragments of a theory are used which makes for the great generality in the following way. According to our account in section 2.5, objects and their types are identified with property clusters, the clusters usually being of modest size. When, for example, a conservation law is invoked, the specific nature of the conserved quantity is often irrelevant, and certainly the full set of properties constituting the domain of the theory at hand is never used. All that matters is the mathematical form of the function and that there is no net loss or gain of the relevant quantity. It is this welcome flexibility of application for the templates that allows us to take the templates as our focus rather than specific theories when investigating computational science.

3.11 The Role of Subject-Specific Knowledge in the Construction and Evaluation of Templates

> In individual cases it is necessary to choose an appropriate sigma-algebra and construct a probability measure on it. The procedure varies from case to case and it is impossible to describe a general method.[61]

There is a curious tension involved in the use of computational templates. As we have seen, computational templates straddle many different areas of science in the sense that many of them can be successfully applied to a wide variety of disparate subject matters. Take the basic idea of fitness

---

[60]It is easy to imagine situations in which this would not be true and in which the Poisson model would not hold, as when cars travel in armed convoys for security reasons.

[61]Feller 1968b, p. 115.

landscapes, a technique pioneered by Sewall Wright, J.B.S. Haldane, and R.A. Fisher. The underlying techniques can be applied to biological evolution, the evolution of economies, and the survival strategies of corporations. The mathematical techniques developed for modeling the simulated annealing of metals were adapted to provide the basis for the revival of neural network research.[62] Spin glass models, originally devised in physics to model solid state phenomena,[63] have been adopted by complexity theorists for a wide range of phenomena, including modeling political coalitions. Cellular automata, invented by von Neumann as a device to represent self-reproducing automata, have been adapted, elaborated, and applied to agent-based models in sociology, economics, and anthropology.

On the other hand, a considerable amount of subject-specific knowledge is needed in order to effectively fit the templates to actual cases. The choice of which force function or probability distribution should be used to bring a theoretical template down to the level of a computational template requires knowledge of the subject matter—the 'theory' does not contain that information and more than inferential skills are needed in the construction of the computational template.

In general, matter constrains method. By this I mean that the intrinsic nature of a given phenomenon renders certain methods and forms of inquiry impotent for that subject matter, whereas other subject matters will yield to those means. Heating plays a key role in polymerase chain reaction processes, but virtually none in studying radioactive decay; randomized controlled trials are the gold standard in epidemiological investigations, but are unnecessary for electromagnetic experiments; careful attention to statistical analysis is required in mouse models of carcinogenesis, but such analysis is often casual window dressing in physics.[64] Subject matter does not determine method—too many other factors enter for that to be true—but the philosophical dream for much of the twentieth century was that logical, or at least formal, methods could, in a subject-matter-independent way, illuminate some, if not all, of scientific method. Although these subject-independent approaches achieved impressive results and research continues in that tradition, there are quite serious limitations on the extent to which those accounts are capable of success.

---

[62]Hopfield and Tank 1986.
[63]See, e.g., Mezard et al. 1987.
[64]For evidence on the last, see Humphreys 1976, appendix.

Consider a specific example that I shall elaborate in some detail to bring out the various ways in which knowledge of a variety of subject matters is needed in the construction and correction of the computational template. It is an important issue for the United States Environmental Protection Agency whether increased particulate matter (PM) concentrations in the atmosphere cause an increase in human mortality and morbidity. Particulate matter comes from a variety of sources, such as diesel exhaust, industrial smoke, dust from construction sites, burning waste, forest fires, and so forth. It comes in various sizes, and it is conjectured that smaller particles (less than 10 μm in diameter) are a greater health hazard than are larger particles. The current air quality standard is that the concentration of such particles, denoted by PM_{10}, should not go above $150 \, \mu g/m^3$ averaged over a 24-hour period. Suppose we want to arrive at a model that will allow us to represent the causal relations between ambient levels of PM_{10} and morbidity in order to predict the latter. There are some obvious subject-specific choices that have to be made before constructing the model: How do we measure morbidity? Some effects of PM_{10} exposure are deaths due to acute cardiopulmonary failure, emergency room admissions for treatment of asthma, and reduced respiratory function. What is the appropriate time lag between exposure and effect? The usual choice is 0–4 days. Which potential confounders should be controlled for? Temperature affects the onset of asthmatic episodes, as does the season; flu epidemics lead to increased hospital admissions for pulmonary distress; and the copollutants sulfur dioxide, carbon monoxide, and ozone, which are by products of PM emissions, can also affect respiratory health.

There are less obvious items of knowledge that must be used, some of which are 'common sense' and others of which are more subject specific, but ignorance of which can easily ruin our evaluation of the template. PM levels in cigarette smoke are many times higher than the ambient levels in air, and so can swamp the effects of exhaust emissions. This would seem to present a straightforward case for controlling for cigarette smoking in the model, except that chronic obstructive pulmonary disease, one of the measures of morbidity, is not exacerbated by smoking in many individuals. Perhaps secondhand smoking could be a problem, for smoking a pack of cigarettes raises the ambient PM levels by $20 \, \mu g/m^3$. Yet if the individual is in a fully air-conditioned building, the ambient levels are lowered by $42 \, \mu g/m^3$, thus complicating the issue. Air quality monitors are located outdoors, yet on average people spend more than 90 percent of their time indoors. In the elderly, it is even more. The air quality meters, being scattered, are not representative of local levels in many cases, and educated interpolations must be made. The time lag for admissions for treatment of asthma is about 41 hours, close to the upper bound of many of the time series correlations, and so a number of such admissions related to PM will occur too late to be caught by such studies. Finally, ecological inferences made from aggregate data on exposure to individual exposure levels are frequently invalid.[65] It is thus not only impossible (or perhaps irresponsible) to assess a given model of the links between PM levels and morbidity without a considerable amount of subject-specific knowledge being brought to bear on the assessment, it is also quite clear that adjustments to defective models will be made not on grounds of convention but on the basis of informed knowledge of the area, with a strong sense of the rank order of the defects.

There do exist cases in which the evidence in favor of a causal connection between variables is so strong that the models are robust under quite significant epistemic defects, but this is unusual. For example, Doll and Peto's epidemiological work that established a causal link between smoking and lung cancer was accepted, despite the absence of any widely accepted mechanism linking smoking and lung cancer, because of the highly invariant nature of the link across populations and contexts.[66]

Such examples, which are unusual, do not undermine the general point that a high level of subject-specific knowledge is required to justify causal inferences.[67] The need for such knowledge means neither that

[65]These and other cautions can be found in Abbey et al. 1995; Gamble and Lewis 1996; and Gamble 1998, among others.

[66]A different example comes from traditional physics. Here is what one formerly standard text says about using Laplace's equation to model electrostatic phenomena: "If electrostatic problems always involved localized discrete or continuous distribution of charge with no boundary surfaces, the general solution (1.17) [i.e., the simple integration of the charge density over all charges in the universe] would be the most convenient and straightforward solution to any problem. There would be no need of Poisson's or Laplace's equation. In actual fact, of course, many, if not most, of the problems of electrostatics involve finite regions of space, with or without charge inside, and with prescribed boundary conditions on the bounding surfaces. These boundary conditions may be simulated . . . but (1.17) becomes inconvenient as a means of calculating the potential, except in simple cases. . . . The question arises as to what are the boundary conditions appropriate for Poisson's (or Laplace's) equation in order that a unique and well-behaved (i.e. physically reasonable) solution exist inside the bounded region. Physical experience leads us to believe that specification of the potential on a closed surface . . . defines a unique potential problem. This is called a Dirichlet problem" (Jackson 1962, pp. 14–16). So again 'physical experience'—subject-matter-specific knowledge—lies behind the use of a standard model (and enhancing it for more complex applications).

[67]This fact is explicitly acknowledged in Pearl 2000.

general conclusions about science are impossible nor that general philosophy of science is ill conceived. Philosophy of science does not reduce to case studies because the morals drawn here about computational science transcend the particular cases considered. In a similar way statistics, which also has ambitions to great generality, requires subject-specific knowledge to be applied in a thoughtful way, but it has, nonetheless, been quite successful in providing general methodological maxims.

3.12 Syntax Matters

> The chasm between naturalist and mathematician exists to a large extent because they view theory in different ways: the naturalist is interested in the premises *per se*, and in what they say about the biological system; the mathematician is accustomed to being presented with a set of axioms and to working entirely within the world thereby defined. The mathematician must learn that the biologist's fundamental interest is in the truth or falsity of the axioms; the biologist must appreciate the power of the deductive phase. That the conclusions the mathematician derives are implicit in the assumptions is a trivial remark: Even exceedingly simple postulates may carry implications that the finest human mind cannot grasp without the aid of mathematical formalism.[68]

Templates have multiple functions. One of these functions is to represent the world; another is to facilitate calculations. It is unfortunate that twentieth-century philosophy of science has focused so heavily on the first of these and has largely relegated the second function to the end of deducing consequences from the theory. The functions of accurate representation and ease of calculation are not always simultaneously optimizable, and in some cases the computational demands are at odds with the representational aspect.[69] The Navier-Stokes equations are generally regarded as the standard equations representing turbulent flow and have been used as the basic representational device in that area since the nineteenth century. Yet computational procedures are currently effective

---

[68]Levin 1980, p. 423.

[69]This is also true of traditional deductions. Economical representational apparatus can lead to significant difficulties in deducing consequences from those representations. For some examples, see Humphreys 1993a. Those issues are, once again, obscured when the "in principle" attitude is adopted.

in accurately applying the Navier-Stokes representations only at low Reynolds numbers, and the primary goal of research since the 1980s has been to achieve a steady improvement in computational efficiency in order to improve the detail of applications. In the other direction, transformations of representations are commonly used in theories to facilitate computation, but many of these take us away from the realm of reality into the realm of the fictional.

Philosophical conceptions of theories that are concerned only with the issue of how the theory is applied in principle do not address these problems. Two such conceptions of theories dominated the philosophy of science in the twentieth century, the *syntactic* account and the *semantic* account. We saw the basic outlines of the syntactic approach in section 3.3. It views theories as deductively organized sets of sentences in some precisely specified language, often first-order logic, although this is a poor choice, given the inability of first-order logic to represent such core mathematical concepts as 'has measure zero' and 'random variable'.[70] The nonlogical terms in the language, which are initially uninterpreted and thus are purely syntactic objects, are given an interpretation through 'correspondence rules' that link the syntax to objects and properties in the world. Certain sentences are picked out as representing the fundamental principles (axioms) of the theory, and by means of standard deductive procedures on those axioms, testable consequences are derived. If we are lucky, the axiom set will be complete (i.e., all true sentences about the subject matter will follow logically from those axioms). The justification of the axioms is carried out through testing their empirical consequences rather than through some internal or antecedent process of justification. (In contrast, analytically true axioms have an internal justification.) For certain philosophical purposes, primarily concerning very general and fundamental epistemological problems, this picture can be used effectively. These syntactic accounts of theories, as traditionally presented, can reconcile representational and computational ends, but only at such a high level of abstraction that the realities of scientific calculation are left behind.

The chief rival of the syntactic approach, what has come to be known as the semantic view of theories, abstracts from the particular linguistic representation of the theory and identifies the theory with its class of

---

[70]See Barwise and Feferman 1985, chap. 1.

mathematical models; that is, the class of set-theoretic structures that make the various linguistic formulations true.[71] A primary reason for this abstractive step is to remove the syntactic peculiarities of different formulations of the same theory and to focus on its underlying structure. Thus, Newton's own formulation of classical mechanics in the *Principia*, modern Lagrangian formulations, and modern Hamiltonian formulations[72] are all considered on this view to be variants of the same underlying theory—classical mechanics.

There are significant virtues to the semantic approach, and it has led to important insights into the abstract structure of theories. Yet when we ask the question "How are theories applied?" the act of abstracting from the particular mathematical representation, an abstraction that is essential to the semantic approach, eliminates exactly the differences that are in practice crucial to how a theory is applied. For many years Patrick Suppes, one of the originators of the semantic approach, urged upon us the slogan "To axiomatize a theory is to define a set-theoretic predicate." That is a powerful and interesting view, but the mathematics used to represent theories ought not to be largely restricted to set theory and should be focused instead upon the grittier theories actually used in applications.[73] The change from rectilinear to polar coordinates, from a laboratory frame to a center-of-mass frame, from a Lagrangian to a Hamiltonian representation, from one representation that does not allow the separation of variables to another that does—all these and more make the difference between a computationally intractable and a computationally tractable representation.

Another advantage of such syntactic transformations is that a successful model can sometimes be easily extended from an initial application to other, similar systems. For example, when using the Schrödinger equation, spherically symmetric solutions for hydrogen can be extended to singly ionized helium, double-ionized lithium, and so on by using a_0/Z

---

[71]There are various versions of the semantic account, such as the state-space approach and the structuralist approach, but again, nothing that I say here depends upon those differences. For examples of the semantic approach see, e.g., Suppes 1970; van Fraassen 1980; Balzer et al. 1987.

[72]See, e.g., Kibble 1966, §3.6 for the second and §13.1 for the third of these.

[73]van Fraassen 1980 represents Suppes as holding that "The right language for science is mathematics, not metamathematics." Suppes (personal communication) has not been able to locate that line in his writings, but reports that he likes the spirit of the remark.

as the distance unit (where a_0 is the Bohr radius and Z is the atomic number) and $Z^2 E_R$ as the energy unit where E_R is the Rydberg energy ($=13.6\,\mathrm{eV}$, the ionization energy of the ground state of the hydrogen atom).[74] In these transformation processes, fictitious objects such as the center of mass are often introduced. This requires us to remain aware of the ways in which computational interests can be at odds with realism and to adopt the position of selective realism described in section 3.8.

The point that representations matter can be reinforced using a simple problem. Consider a man hiking to the top of a mountain along a trail. He begins at the foot of the trail at 6 A.M. on the first day and reaches the top at 6 P.M. He starts down again at 6 A.M. the next day and reaches the bottom at some time later that day. Question: Is there any point on the trail that the hiker occupies at exactly the same time of the day on both the first and the second day? It is not even evident when this puzzle is first presented that it is well posed, but once the right way to represent the problem has been found, the answer is obvious and the basis for the solution is as good a candidate as any for a pure, nonanalytic, a priori justification. (If you are stumped, see footnote 75.)

Consider a second example. You and your opponent are to choose integers from the list 1 through 9. The first player chooses one of those integers, the opponent picks one from those remaining, and so on. The object is to be the first player to get three numbers that add up to 15. Try it with another player and see if you can determine if there is a winning strategy or, less ambitiously, a strategy for not losing. Initially, not so easy.

---

[74] See French and Taylor 1998, p. 176.

[75] Even thinking of solving this by kinematical theory is hopeless; instead, simply use the graphical representation of an imaginary second hiker moving down the trail on the first day. It is immediately evident that the real hiker's and the imaginary hiker's paths cross at some point. This is one example where a visual representation is clearly superior to a syntactic representation. There is a well-known problem that also illustrates the importance of specific representations: Two trains are approaching one another from opposite directions on the same track. Each is moving at 50 m.p.h. and they are initially 25 miles apart. A fly buzzes back and forth between the engines at 4 m.p.h. How far does the fly travel before it is squashed when the engines collide? A possibly apocryphal story about John von Neumann reports that when he was presented with this problem, he instantly replied with the correct answer. The poser, impressed, suggested that he must have seen the trick right away. "Not in the least," replied Von Neumann, "I summed the relevant series." Lesser minds eventually see that the trains travel fifteen minutes before colliding, during which time the fly, traveling at a constant rate of speed, has traveled 1 mile.

But find a different representation and you will immediately see how to do it.[76]

Although these examples are not, of course, scientific theories, they illustrate something long recognized by problem solvers: *To solve a problem, the specific form of the representation matters.* The undoubted virtues of highly abstract representations in the philosophy of science have led to this obvious point being obscured or discounted as being of merely practical interest, and not only in the case of theories. Take the case of data representations. The dominant logical tradition has emphasized propositional representations in the syntactic approach, or set-theoretical representations in the semantic approach, simply because, in principle, any data can be thus represented.[77] Yet this 'in principle' argument ignores the epistemic impossibility of humans' assimilating the information contained in millions of data points from astrophysical or biological instruments when the data are presented propositionally. Whatever the 'rational reconstruction' of scientific practice might mean, any reconstruction of astrophysical practice along propositional lines would be massively irrational for the purpose of understanding how theories are, and ought to be, applied when humans are the intended epistemological agents. (In contrast, many machines would have no use for visual representations.)

To take a last scientific example, in an abstract sense the Heisenberg, Schrödinger, and interactive representations of nonrelativistic quantum theory are variants of the same underlying theory. But in terms of applications, the Schrödinger approach is better suited for some purposes and the Heisenberg approach for others. The abstract approaches to theories place in the same equivalence class formulations of theories that in practice are computationally radically different from one another. And so when we ask "How do we apply theories?," we must be careful that the theory in question is not the philosopher's purified entity but the concrete piece of

---

[76]The trick is to arrange the integers in the form of a 3 × 3 magic square.

$$
\begin{array}{ccc}
8 & 3 & 4 \\
1 & 5 & 9 \\
6 & 7 & 2
\end{array}
$$

Then use the well-known tic-tac-toe strategy and you can't lose if you go first. I learned this example from Stephanie Guerlain of the University of Virginia Systems Engineering Department.

[77]The dominance of such propositional representations has come under question in logic in recent years, with a revival of interest in diagrammatic representations. For examples of these, see Barwise and Etchemendy 1992; and Shin 1994.

syntax upon which the difficult work of application has to be carried out. The pristine objects of our philosophical affections are of enormous interest, no doubt about that, but they can distract us from the real issues.

3.13 A Brief Comparison with Kuhn's Views

Our templates serve one of the functions of what Thomas Kuhn called 'symbolic generalizations'.[78] These are part of the stock-in-trade taught to developing practitioners of a field, and their mastery is considered indispensable for successful practice in that field. Kuhn's account of these tools focuses exclusively on what are often called "off-the-shelf models" because his symbolic generalizations are "developed without questions or dissent by group members, [and] can be readily cast in a logical form $(x)(y)(z)\phi(x, y, z)$."[79] This uncritical attitude toward the templates no doubt accurately describes applications of science to routine matters or during the early stages of learning by a student, but it is not characteristic of the research use of mathematical models that drives scientific progress. For much of that progress involves construction of, or revision of, the 'symbolic generalizations', and the construction and revision processes are certainly not conducted without question or dissent. Indeed, advances in understanding, both of the power and of the limitations of various symbolic generalizations come through a thorough understanding of those two processes.

Kuhn also claims[80] that some symbolic generalizations serve a definitional as well as a lawlike function, and (hence) that "all revolutions involve, amongst other things, the abandonment of generalizations the force of which had previously been in some part that of tautologies."[81] But the case he cites to illustrate this, Ohm's Law, cannot be an example of what Kuhn is claiming. For the 'symbolic generalization' $I = V/R$ that Kuhn uses in support of his claim is constructible from $\mathbf{J} = \sigma\mathbf{E}$, where $\mathbf{J}$ is the current density, $\mathbf{E}$ is the electric field, and σ is the specific conductivity, giving (for a wire of cross section A, length l) $I = A\sigma V/l$. Resistance is defined as $R = l/A\sigma$, not via Ohm's Law, and current I is defined by $I = \oint \underline{J} \cdot d\underline{S} = \oint n e \underline{v} \cdot d\underline{S}$ where $\underline{v}$ is the velocity of particles of charge e with density n per unit volume, again independently of

---

[78]Kuhn 1970, pp. 182–183.
[79]Ibid., p. 182.
[80]Ibid., p. 183.
[81]Ibid., p. 184.

Ohm's Law. Thus Kuhn's claim is, at best, a historical point about the way the definitions of I and R were once conceived, and fails as a claim about conceptual links. Moreover, it is known that σ is a function of temperature, which for true conductors is proportional to T^{-1} and for semiconductors is proportional to $e^{-b/T}$, allowing a refinement of, and hence a change in, the initial model to bring it into better agreement with data. None of these processes are revolutionary in the least. They simply reflect progress in the construction and understanding of 'Ohm's Law' and its replacement with functional forms that reflect improvements in the resulting model.[82]

Kuhn's account of symbolic generalizations in normal science thus says too little about why these generalizations are modified within paradigms in the way they are, when they are. The suggested use of symbolic generalizations reminds one of the kind of simplistic modeling that occurs when existing templates are uncritically applied to phenomena on the basis of some vaguely perceived sense of resemblance to previously successful applications. While there is no doubt that this sort of routine modeling frequently occurs, it is curious that Kuhn, who knew that we cannot learn about science from just its finished products, should be so wedded to textbook methods in this case. The rote learning and use of off-the-shelf models is certainly not characteristic of the cutting-edge use of research models, even within normal science.

The examples of constructions we have seen do illuminate one of Kuhn's insights—his use of the term 'exemplar'.[83] In this use, an exemplar is a stereotypical application of a theory (or symbolic generalization), typified by the kinds of 'worked examples' often found at the end of chapters in physics textbooks. In such ideal cases of the theory—for example, a ladder leaning against a wall under the action of gravity with friction at both ends and no other forces acting upon it—we have a stereotypical 'clean' system to which the equations of statics can be uncontroversially applied. Having mastered such exemplars, the student is then expected to transfer the applications to other systems that resemble the exemplar more or less closely. There are thus three skills that must be learned: applying the template to the exemplar, recognizing that some other systems are similar in relevant respects to the exemplar, and

---

[82]This is not the only place where Kuhn made incorrect claims about the defined status of certain concepts. See Humphreys 1993b; and Cargile 1987 for arguments to that effect.

[83]See Kuhn 1970, pp. 186ff., 192, 200.

modifying the template to fit the new example. This is the important use of 'exemplar' to which Kuhn drew our attention, and it was an important insight into the role of examples in science.

However, Kuhn goes on to say: "The practice of normal science depends on the ability, acquired from exemplars, to group objects and situations into similarity sets which are primitive in the sense that the grouping is done without an answer to the question 'Similar with respect to what?' "[84] As his quote indicates, one of Kuhn's principal points about exemplars was that science is not applied by following rules.[85] This is too crude. Systems can be grouped according to whether they can plausibly be seen as conforming to the same equations, and this plausibility is based not upon some primitive similarity but upon whether the systems' structure is such that it satisfies the construction assumptions leading to the computational template, the assessment of which can be, and often is, explicit and objectively debatable. For example, there is in certain cases a rule for applying the Poisson model. Each of the four assumptions underlying the Poisson model can be inductively justified by an explicit rule based upon observations of finite frequencies. There will, of course, be room for disagreement about, say, the independence of disjoint intervals, but this disagreement can be adjudicated by appeal to sample sizes and other criteria. There is no guarantee that the resulting model will be an accurate description of the phenomena to which it is applied, but this is true of any a posteriori process for applying theories to reality. The grouping is thus not primitive, but can be justified in these structural terms. Kuhn's image suggests that we examine the systems as unanalyzable entities and then decide whether they are similar to the exemplar, but this is quite misleading.

3.14 Computational Models

We have discussed at some length the concept of a template and features associated with it. We can bring together these features by introducing the concept of a computational model. A computational model has six components:

1. A computational template, often consisting of a differential equation or a set of such, together with the appropriate boundary or

---

[84]Ibid., p. 200.
[85]Ibid., pp. 191–192.

initial conditions types. Integral equations, difference equations, simultaneous equations, iterative equations, and other formal apparatus can also be used. These syntactic objects provide the basic computational form of the model.

2. The construction assumptions.
3. The correction set.
4. An interpretation.
5. The initial justification of the template.
6. An output representation, which can be a data array, a graphical display, a general solution in the form of a function, or a number of other forms. The output representation plays a key role in computer simulations, as we shall see in chapter 4.

The sextuple <Template, Construction Assumptions, Correction Set, Interpretation, Initial Justification, Output Representation> then constitutes a *computational model*, which can be an autonomous object of study. If we consider only the formal template, there is no essential subject dependency of the models, but as the assumptions, correction set, interpretation, and justification are specified, we recover some of the more traditional, subject-specific, organizational structure of the sciences that is lost with the versatility of the templates. With these models, scientific knowledge is contained in and conveyed through the entire sextuple—the formal template is simply one component among many. Although each component in the sextuple can be the object of focused attention, it will be rare that it can be evaluated without considering at least some of the remaining components.

Often, the construction of a template that is not computationally tractable will precede its use in a computational model, sometimes by a considerable period of time. For example, in 1942, Kolmogorov developed a now widely used two-equation model of turbulence,[86] but it was not until the 1970s that the model could be applied, when computational methods were finally developed to implement the equations. It is also common for a reciprocal process of adjustment between protomodels and simulation runs to result in the development of a properly articulated computational model. The links that have to be constructed between a computational model and its consequences as represented by the solution space are not always just "more of the same" in the sense that it requires only an ingenious application of preexisting rules of inference, as the

[86]Kolmogorov 1942.

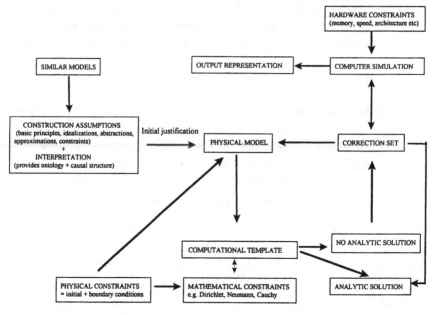

Figure 3.1

standard picture of axiomatically formulated theories would lead us to believe. The task will ordinarily be composed of a complex and interconnected set of factors including (1) the development of concrete computational devices of sufficient power; (2) the justification, and occasionally the development, of approximation methods; (3) the construction of algorithms and programs to implement (2); and (4) idealizations. We can represent the stages in the construction and refinement of computational models by figure 3.1. (To preserve clarity, links between an analytic solution and a simulation based upon that solution have been omitted.)

4

Computer Simulations

4.1 A Definition

> The acceptance by medicinal chemists of molecular modeling
> was favored by the fact that the structure-activity correlations are
> represented by 3-D visualizations of molecular structures and not
> by mathematical equations.[1]

Because our goal is to establish the various ways in which computational
science involves new scientific methods, it will be helpful to examine a
representative area that most researchers would agree falls under the
heading. In this chapter I shall focus on the important subarea of computer simulations. When discussing these activities, we must be aware of a
number of things. The entire field of computer simulations, like computational science, is relatively new and is rapidly evolving. Unlike well-
entrenched fields, techniques that are now widely used may well be of
minor interest twenty years hence as developments in computer architecture, numerical methods, and software routines occur. Moreover, some
of the methods are highly problem-specific. Methods that work for one
kind of simulation are often not applicable to, or work less efficiently for,
other simulations, and arguments to general conclusions are thus doubly
hard to come by. For these reasons, I shall discuss only general issues that
apply to a broad variety of approaches and which seem to have more than
a temporary status.

We begin with a simple informal definition of the general category of
computational science: *Computational science* consists in the development, exploration, and implementation of computational models (in the

[1]Cohen 1996, p. 4

sense given in section 3.14) of nonmathematical systems using concrete computational devices.

At the level of this informal definition, computational science has been around for a very long time indeed. Humans are concrete computational devices, so the calculation of the motion of Venus by ancient Greek astronomers using Ptolemaic theories counts as computational science and so does the computing of projectile trajectories with the aid of mechanical hand-operated calculators.[2] In the 1940s, the modeling of nuclear explosions by some of the earliest electronic computers constituted the cutting edge of computational science. Nowadays, there is an entire range of sophisticated computational aids, from student software packages for computational chemistry to the most demanding climate modeling carried out on the latest supercomputer. There is a continuum of extrapolation here, just as there is a continuum in extending our observational powers from the simplest optical microscopes to the currently most powerful magnifying devices.

The simple definition given above does not, however, fully capture what is distinctive about contemporary computational science, even though it does cover an enormous variety of methods falling into the domain of computational science. These include modeling, prediction, design, discovery, and analysis of systems; discrete-state and continuous-state simulations; deterministic and stochastic simulations; subject-dependent methods and general methods; difference equation methods; ordinary and partial differential equation methods; agent-based modeling; analog, hybrid, and digital implementations; Monte Carlo methods; molecular dynamics; Brownian dynamics; semi-empirical methods of computational chemistry; chaos theory; and a whole host of other methods that fall under the loose heading of computational science. Any attempt to capture these various methods under a single precise definition would be a daunting, if not impossible, task. Moreover, much of computational science falls into

---

[2]When I was a schoolboy, a very elderly gentleman came to our science class to talk. He seemed weighed down by the problems of old age. The topic of his talk was the development during World War I of forward-mounted machine guns that could fire through the propellers of aeroplanes, the synchronization mechanisms of which were adapted from mechanical calculators. As he regaled us with stories of miscalculations that resulted in pilots shooting off the propellers of their own planes and plummeting to their deaths, he began to pull from his jacket pockets various gears, levers, and springs, all machined from steel and brass. As his talk progressed, he became increasingly more sprightly and cheerful, the talk ending with a large pile of metal on the table and the speaker considerably less bent.

the realm of computer-assisted science, within which computational tools are used only as a supplement to more traditional methods. So I shall restrict my precise definition to the category of computer simulations, which forms a special but very important subdomain of computational science.[3]

Simulations are widely used to explore mathematical models that are analytically intractable. But they are also used when, for practical or ethical reasons, numerical experiments are more appropriate than empirical experiments. For example, some empirical experiments, such as intervening in the U.S. economy, are too costly, and hence are replaced by simulations; some are too uncertain in their outcomes at various stages of theoretical research, so that such things as the containment of controlled fusion in a magnetic bottle are better simulated than created; some are too time-consuming, such as following the movement of real sand dunes, so that simulations are used to compress the natural time scale; others are politically unacceptable, so that nuclear explosions are simulated rather than investigated empirically by setting off real, dirty devices; or they are ethically undesirable, so that we protect amphibians by replacing the dissection of real frogs with simulated dissections or the diffusion of drugs in a community is simulated by using epidemiological models. Practically impossible experiments include rotating the angle of sight of galaxies, reproducing the formation of thin disks around black holes, investigating oil reservoir flows, earthquake modeling, and many others. In other cases simulation is used as a pre-experimental technique, when trial runs of an experiment are performed numerically in order to optimize the design of the physical experiment.

We can approach the task of defining computer simulations by looking at the difference between a representation and a simulation. Simulations rely on an underlying computational model or are themselves models of a system, and hence either involve or are themselves representations. So what is the relation between a computational model, a simulation, and a representation? In the first article that I wrote on simulations, I left open what counted as a simulation and used what I called a 'working definition': "A computer simulation is any computer-implemented method for exploring the properties of mathematical models

[3]One of the first philosophers to have discussed simulations explicitly is Sayre 1965. (See also Crosson and Sayre 1963.) As the importance of this field becomes recognized among philosophers, I hope that his priority in this area will be properly recognized.

where analytic methods are unavailable."[4] This working definition was both too narrow and too broad. In an insightful article,[5] Stephan Hartmann has argued that the definition is too narrow, because there are many simulations of processes for which analytically solvable models are available.[6] The working definition is also too broad because it covers areas of computer-assisted science that are not simulations. In addition to his critical points, Hartmann provided the basis for what now seems to me to be the correct definition of computer simulations. Hartmann's own position needs some revision, but his principal insight is correct. He writes:

> A model is called *dynamic*, if it . . . includes assumptions about the time-evolution of the system. . . . Simulations are closely related to dynamic models. More concretely, a simulation results when the equations of the underlying dynamic model are solved. This model is designed to imitate the time evolution of a real system. To put it another way, *a simulation imitates one process by another process.* In this definition, the term "process" refers solely to some object or system whose state changes in time. If the simulation is run on a computer, it is called a *computer simulation.*[7]

It is the idea of a temporally dynamic process that is key here, although we must leave room for simulations of static objects as well. A specific example will help. Suppose that we are computationally simulating the orbital motion of a planet. This process will consist in successive computations of the planet's state (position and velocity) at discrete time intervals, using the mathematical equations that constitute a model of the orbital kinematics, together with an arrangement of the outputs from these computations into a representation of the planet's motion. In this particular example, it is the behavior of the system that is being simulated and there is no explicit concern about whether the processes which generate the behavior are accurately modeled.

---

[4]Humphreys 1991, p. 501.

[5]Hartmann 1996.

[6]In correspondence with Fritz Rohrlich and Ron Laymon dated June 29, 1990, I wrote "I'm not sure that we ought to exclude as a simulation some computer run that analytically works out the trajectory of a falling body in a vacuum under a uniform gravitational force, for example." Eventually, I settled in the published paper for a narrower conception, wrongly as it turned out.

[7]Hartmann 1996, p. 82.

The core of this simulation is the working out of the successive positions and velocities of the planet. To capture the special nature of computer simulations, we need to focus on the concrete devices by means of which the computational science is carried out. Included in this class of computational devices are both digital and analog computers; that is, analog devices that are used for computational purposes and not just as physical models. I shall, in contrast, set aside here noncomputational simulations such as occur when a scale model of an airframe is used in a wind tunnel test. It is important that the simulation is actually carried out on a concrete device, for mere abstract representations of computations do not count as falling within the realm of computer simulations. This is because constraints occur in physical computational devices that are omitted in abstract accounts of computation, such as bounded memory storage, access speed, truncation errors, and so on, yet these are crucially important in computer simulations.[8] Furthermore, the whole process between data input and output must be run on a computer in order for it to be a computer simulation.

It is the fact that this solution process is carried out in real time even when massively parallel processing is used which crucially differentiates it from the computational model that underlies it, for logical and mathematical representations are essentially atemporal. The move from abstract representations of logical and mathematical inferences to temporal processes of calculation is what makes these processes amenable to amplification and allows the exploitation of our two principles of section 3.2. The move also entails that none of the representational categories of section 3.3 — syntactic or semantic theories, models, research programmes, or paradigms — are able to capture what simulations can do. I shall call this temporal part of the computational process the *core simulation*. Each individual computation in the core simulation could, from the 'in principle'

---

[8]There is a different use of the term 'simulation', which is standard in computer science, that is closely tied to the concept of a virtual machine. "A virtual machine is a 'machine' that owes its existence solely to a program that runs (perhaps with other intervening stages) on a real, physical machine and causes it to imitate the usually more complex machine to which we address our instructions. Such high level programming languages as LISP, PROLOG, and POPII thus define virtual machines" (Clark 1989, p. 12). This is not the sense in which the term 'simulation' is to be taken here, although there are connections that can be made, and the sense just discussed is perhaps better called 'emulation' rather than 'simulation'. Nor do I intend this account to cover what is called 'simulation theory' in cognitive science, wherein understanding of others comes through projectively imagining ourselves in their place.

perspective, be viewed as a replacement for the human calculations that are usually infeasible in practice, if not individually then sequentially. In our planetary example, this core simulation leads to specific numerical values of position and velocity. The computational process is thus an amplification instrument—it is speeding up what an unaided human could do—but this amplification takes the method into an extended realm in a way which closely resembles the way that moving into the realm of the humanly unobservable does for instruments. We shall explore some of these similarities in section 4.3.

It is with the representation of the output that one of the key methodological features of simulations emerges. Consider three different ways of representing the results of the core simulation. In the first way, the results are displayed in a numerical table. In the second way, the results are represented graphically by an elliptical shape. In the third way, the output is displayed dynamically via an elliptical motion. If the results of the core simulation are displayed in a numerical array, that array will be a *static representation* of the planet's motion, as will be a display in the form of a static ellipse. If the results are graphically displayed as an elliptical motion, it will constitute a *dynamical representation*.

We can now formulate the following definition: System S provides a core simulation of an object or process B just in case S is a concrete computational device that produces, via a temporal process, solutions to a computational model (in the sense of section 3.14) that correctly represents B, either dynamically or statically. If in addition the computational model used by S correctly represents the structure of the real system R, then S provides a core simulation of system R with respect to B.

I note that the definition covers cases in which S is a human calculator as well as cases in which S is an artificial computational device, even though human simulations are too slow to be of much interest in practice. We thus preserve the parallels between extensions of our perceptual powers and extensions of our computational powers. Simulations of a static system or of static behavior—for example, that of a system in equilibrium—can be viewed as special cases of core simulations, in part because successive states need to be calculated to simulate the unchanging state of the system.[9] It has also become acceptable to speak of simulating certain kinds of mathematical objects, such as probability

---

[9]Simulations of boundary value problems are frequently concerned with static phenomena, as in applications of Poisson's equation $\nabla^2 u(x, y, z) = \rho(x, y, z)$.

distributions. Where the object of the simulation is a purely mathematical object, the idea of the object being generated by a system will not apply and so only the first part of the definition comes into play, with B, in these cases, being the mathematical object. B will, unless time-dependent, have a static representation.

In order to get from a core simulation to a full simulation, it is important that the successive solutions which have been computed in the core simulation are arranged in an order that, in the case of our example, correctly represents the successive positions of the planet in its orbit. If the solutions were computed in a random order and a numerical array of the solutions presented them as such, this would not be a simulation of the planet's orbital motion but, at best, of something else. In contrast, if the solutions were computed in parallel, or in an order different from the order in which they occur in the actual orbit, but were then systematically reordered so as to conform to the actual order in the output representation, we would have a simulation of the orbital behavior, although not of the planetary system because the underlying computational processes misrepresent the dynamics of the real system. So the process that constitutes the simulation consists of two linked processes—the core process of calculating the solutions to the model, within which the order of calculations is important for simulations of the system but unimportant for simulations of its behavior, and the process of presenting them in a way that is either a static or a dynamic representation of the motion of the real planet, within which the order of representation is crucial. The internal clock in a computer simulation is ordinarily crucial for arriving at the correct ordering in core simulations of systems and in the representations of their behavior, although some early computers did not use them, relying on alternative means of ordering their sequence of operations. Nevertheless, even in those cases, the computation must be carried out in real time, as opposed to being conceived of abstractly.

When both a core simulation of some behavior and a correct representation of the output from that core simulation are present, we have a full computer simulation of that behavior.[10] And similarly, mutatis mutandis, for a full simulation of a system.

---

[10]'Correct' includes having a dynamic representation for dynamic behavior and a static representation for static behavior. There is a danger of semantic confusion arising in this area. Sometimes the output of a simulation is displayed, and it is said, "That's a simulation of X." But that is simply shorthand for referring to the combined process of core simulation and output representation we have just described.

In the early days of computer simulations, static numerical arrays were all that was available, and it would seem unreasonable to disallow these pioneering efforts as simulations. Nevertheless, the difference between the two types of output turns out to be important for understanding why simulations have introduced a distinctively different method into scientific practice. For the definitions we have given up to this point might seem to give us no reason to claim that computer simulations are essentially different from methods of numerical mathematics. Numerical mathematics is the subject concerned with obtaining numerical values of the solutions to a given mathematical problem; numerical methods is that part of numerical mathematics concerned with finding an approximate, feasible solution to a given problem; and numerical analysis has as its principal task the theoretical analysis of numerical methods and the computed solutions, with particular emphasis on the error between the computed solution and the exact solution.

We cannot identify numerical mathematics with computational simulations, even though the former plays a central role in many areas of the latter. The reasons are twofold. First, computational models are an essential component of simulations, whereas there need be no such model in numerical mathematics. Second, there is a concrete computational element to simulations that need not be present in numerical mathematics. A separate but important issue concerns the place of computer simulations in nonautomated science. A purely automated science of the kind discussed in section 1.2 would attribute no special status to the form of the output representation. It is only when we focus on humans as scientists that the importance of the output representation is apparent, because the information contained in vast numerical arrays is cognitively inaccessible to humans until a conversion instrument is available.

Approximately 60 percent of the human brain's sensory input comes from vision.[11] Because the human eye has a nonlinear response to color, brightness and color mainpulation can often be used in displays to bring out features that might be hidden in a 'natural' image. Similarly, gray-scale images often provide a better representation of structure than does a colored image, as do color and contrast enhancement and edge detection software. It is partly because of the importance for humans of qualitative features such as these that computer visualizations have become so widely used. Scientists have relied on graphical models in the past—recall the

---

[11]Russ 1990, p. 1.

famous photograph of Watson and Crick in front of the molecular model of DNA and the elementary school representation of a magnetic field by iron filings—but visual representations have usually been downplayed as merely heuristic devices. Yet graphical representations are not simply useful; they are in many cases necessary because of the overwhelming amount of data generated by modern instruments, a fact we identified in section 1.2 as the quantity-of-data issue. A flight simulator on which pilots are trained would be significantly lacking in its simulation capabilities if the 'view' from the 'window' was represented in terms of a massive numerical data array.[12] Such data displays in numerical form are impossible for humans to assimilate, whereas the right kind of graphical displays are, perceptually, much easier to understand. To give just one example, a relatively crude finite difference model of the flow of gas near a black hole will, over 10,000 time steps, generate a solution comprising 1.25 billion numerical values.[13] This is why the results of simulations are often presented in the form of animations or movies rather than by static photographs, revealing one limitation of traditional scientific journals with respect to these new methods. Such dynamic presentations have the additional advantage of our being able to see which structures are stable over time. This is also true of many agent-based models—if we display the results of such models numerically or statically, we cannot 'see' the higher-order emergent patterns that result from the interactions between the agents. For the purposes of human science the output representation is more than a heuristic device; it becomes a part of the simulation itself. It is not the extension of human computational abilities alone that produces a new kind of scientific method, but the combination of that extension and a conversion of the mode of presentation.

The appeal to perceptual ease may disturb philosophical readers who might be inclined to identify the simulation with the equivalence class of all simulations that have the same core simulation. We have also become used to arguments in which it is routinely assumed that transformations of graphical representations into propositional representations and vice versa can be performed. Because of the greater generality of propositional representations, they are taken to be the primary object of study. There is, of

---

[12]Issues involving human/data interfaces constitute an important area of research. It has been said that if Apollo XIII had been equipped with monitors which displayed critical data in a more easily assimilated form, the near disaster to that mission would have been much easier to deal with.

[13]Smarr 1985, p. 404.

course, a sense in which this view is correct, because it is the computer code that causes, via the hardware, pixel illumination. Sometimes these outputs are in the shape of formal symbols (formal languages rely only on the geometrical shape of the symbols for individuation) and sometimes they are in the shape of an image. But this transformational view once again concedes too much to an 'in principle' approach. Because one of the goals of current science is human understanding, how the simulation output is represented is of great epistemological import when visual outputs provide considerably greater levels of understanding than do other kinds. The output of the instrument must serve one of the primary goals of contemporary science, which is to produce increased understanding of the phenomena being investigated. Increases in human understanding obviously are not always facilitated by propositional representations, and in some cases are precluded altogether. The form of the representation can profoundly affect our understanding of a problem, and because understanding is an epistemic concept, this is not at root a practical matter but an epistemological one.[14]

4.2 Some Advantages of Computer Simulations

Even given the definition we have formulated, simulation remains a set of techniques rather than a single tool. These techniques include the numerical solution of equations, visualization, error correction on the computational methods, data analysis, model explorations, and so on. The systems that are the subject of simulations need not be complex either in structure or in behavior. As we have established, mathematical intractability can affect differential or integral equations having a quite simple mathematical structure. The behavior of such systems is often not unduly complex, but is merely hard to predict quantitatively without numerical techniques.

Some further remarks may be helpful. Although the everyday use of the term 'simulation' has connotations of deception, so that a simulation has elements of falsity, this has to be taken in a particular way for computer simulations. Inasmuch as the simulation has abstracted from the material content of the system being simulated, has employed various simplifications in the model, and uses only the mathematical form, it

---

[14]I am very much indebted to David Freedman for pressing me to clarify this section. His help in no way implies endorsement of the claims.

obviously and trivially differs from the 'real thing', but in this respect, there is no difference between simulations and any other kind of theoretical model. It is primarily when computer simulations are used in place of empirical experiments that this element of falsity is important. But if the underlying mathematical model can be realistically construed (i.e., it is not a mere heuristic device) and is well confirmed, then the simulation will be as 'realistic' as any theoretical representation is. Of course, approximations and idealizations are often used in the simulation that are additional to those used in the underlying model, but this is a difference in degree rather than in kind.

Because of their increased computational power, simulations usually allow a reduction in the degree of idealization needed in the underlying model, idealizations that often must be made in analytical modeling to achieve tractability. Because they are free from many of the practical limitations of real experiments, simulations and numerical experiments allow more flexibility and scope for changing boundary and initial conditions than do empirical experiments. With the Lotka-Volterra equations (see section 3.4), for example, one can start with many different levels of the initial populations of predator and prey and explore how the population dynamics change with the initial settings rather than being limited by whatever population sizes happen to exist naturally. Computer simulations, unlike empirical experiments, are precisely replicable, even in stochastic simulations. The experiment can be repeated exactly, as many times as we wish, with guarantees of precision that are not possible in the empirical case. Somewhat differently, when dealing with probabilistic outcomes, repeated outcomes differ according to the underlying probability distribution and not because of fluctuations in the underlying experimental conditions.

Many parameters can be varied in simulations that could not be altered in real experiments, perhaps because such a change would violate a law of nature, perhaps because in the case of astrophysical systems the very idea of large-scale experiments is absurd. Indeed, many simulations are examples of what would have been, in technologically more primitive times, thought experiments, as when one removes a planet from a planetary system to see how the orbits of the other planets are affected. The enormous flexibility and precision of simulation methods provide an opportunity to implement Gedankenexperimente in contexts providing much greater precision than is possible with traditional mental implementations, and they are free from the psychological biases that can affect even simple thought experiments. Here we see yet another way in which simulations can occupy the ground between theory and thought experiments on the one hand and traditional experiments on the other, with the ability to violate nomological constraints constituting a key difference from the latter. The ability to change the inverse-square law of gravitation to an r^{-a} form ($a \neq 2$) and show that orbits of the planets become unstable is routine, using simulations. Finally, with simulations you have a guarantee (modulo programming errors) that the assumptions of the underlying model have been explicitly satisfied, a guarantee that, in the case of field experiments especially, is often very hard to acquire.

Simulations can also be used for exploratory studies. A special case of this is when they lead to discoveries that are later derived analytically. For example, the counterrotating vortex structures discovered in slender tori by simulations[15] preceded the development of an analytic solution for these 'planets'.[16] Finally, in evolutionary simulations, altering the actual features of agents or the laws that govern them can show us why certain conditions were necessary in order for certain evolutionary characteristics to have been selected.

4.3 The Microscope Metaphor and Technological Advances

> Although, for many of us, the computer hardware is the least important aspect of what we do, these four phases [of molecular modeling] are defined by the development of computer technology: unless the relevant technology is available, the software and, most importantly, the techniques, cannot flourish.[17]

I have argued that computational devices are the numerical analogues of empirical instruments: They extend our limited computational powers in ways that are similar to the ways in which empirical instruments extend our limited observational powers. While recognizing that such metaphors are potentially misleading, the parallel is sufficiently promising that it is worth pursuing. The specific metaphor which I shall use is that computational methods can be compared to a microscope of variable power. This metaphor is not new, and one occasionally sees it mentioned in the simulation

---

[15]Hawley 1987.
[16]These were later developed by Goodman et al. 1987.
[17]Hubbard 1996, p. 21.

literature, invariably as a throwaway. My purpose here is to relate this metaphor to the more general epistemological issues of chapter 2.

With microscopes, we employ the Overlap Argument in calibrating the instrument by first observing features for which the structure is known via some other mode of access and then requiring that the instrument reproduce that structure. In the case of computational devices, the calibration standards that must be reproduced are analytically derived results which serve as mathematical reference points. With many combinatorially complex calculations we have no cognitive access to the result independent of the simulation itself, and comparison with existing, verified techniques is crucial. In hydrodynamical simulations, for example, analogical reasoning from the behavior of eddies that appear on grids of feasible resolution is extrapolated to the behavior of eddies which are smaller than grid size and hence cannot be represented explicitly. In addition, a technique known as transport modeling can be used within which specific theories are constructed about how small fluctuations in the fluid carry energy, and these local models are used to replace the global Navier-Stokes equations. These techniques are then compared with full numerical simulations at low Reynolds numbers or with observations from real turbulent flows to ensure that matching results are obtained on the calibration models.

This cross-checking against existing methods is not new, for many approximation methods in science are validated against existing approximations[18] and the results from individual human senses are often cross-validated with the results from other senses; the visual sense often, although not invariably, is taken as the key reference standard. All of this reflects use of the overlap argument pattern. Comparison with empirical data from other instruments is also crucial in that the simulation output must conform to whatever such data are available, as are convergence results and stability of the solution under small differences in input variable values.[19]

It is preferable in both the microscope case and the simulation case that the workings of the instrument are known, as we argued in

[18]As in, for example, "The simple ladder approximation is not a good approximation to the eikonal approximation" (Humphreys 1976, p. 137).

[19]There are also technical issues of stability in the approximations used, particularly whether errors expand through iteration. For example, the Crank-Nicholson implicit method is unconditionally stable, whereas the leapfrog method is unstable. Details of such matters can be found in Richtmeyer and Morton 1967.

section 2.8, so that errors introduced by the instrument can be corrected. Such knowledge-based precautions against error are routine with computational methods. For example, a common source of misleading results is the truncation error, generated when numerical representations are stored only to within a set number of significant digits. Thus, using MATLAB release 12 to compute the value of $[1-(1-10^{-17})]$ gives the result 0, whereas the same program computes the value of $[1-(1-10^{-16})]$ as 1.1102×10^{-16}. When used in further computations with suitable multiplying factors, the difference between the two values can produce significant errors.[20] Ignorance of how the algorithm works in such cases can be scientifically disastrous.

As with microscopes, a desirable property of computer simulations is *resolving power*—the greater the degree of resolution, the more detail the simulation provides. By decreasing the grid size (time step or spatial) in finite difference methods, or by increasing the sample size in Monte Carlo or agent-based methods, increased detail about the solution space can be obtained. Because of the quantity of data available, by using graphical representations for the solutions one often can literally see more detail with increased computational power.[21]

This issue of the degree of resolution is common in the sciences. For example, in climatology the most detailed atmospheric models currently have a horizontal resolution only of several hundred kilometers.[22] Here is a representative description of the problem:

> In practice, computing limitations do not allow models of high enough resolution to resolve important sub-grid processes. Phenomena occurring over length scales smaller than those of the most highly resolved GCMs, and that cannot be ignored, include cloud formation and cloud interactions with atmospheric radiation; sulphate aerosol dynamics and light scattering; ocean plumes and boundary layers; sub-grid turbulent eddies in both the atmosphere and oceans; atmosphere/biosphere exchanges of mass, energy and momentum; terrestrial biosphere

---

[20] I owe the example to David Freedman. Hammer et al. 1995 is a good resource for methods that automatically verify whether the results of certain numerical computations are correct.

[21] Well-known examples of this come from fractals and chaos theory. More and more details of Julia sets or of period doubling can be seen as the level of analysis is raised in these two cases.

[22] Harvey et al. 1997, p. 3.

growth, decay and species interactions; and marine biosphere ecosystem dynamics—to cite a few examples. Mismatches between the scale of these processes and computationally realizable grid scales in global models is a well-known problem of Earth system science. To account for sub-grid climate processes, the approach has been to "parametrize"--that is, to use empirical or semi-empirical relations to approximate net (or area-averaged) effects at the resolution scale of the model. . . . It is important to stress that all climate system models contain empirical parametrizations and that no model derives its results entirely from first principles.[23]

This loss of detail due to low resolving power is common, but it does not preclude significant inferences being drawn. For example, paleontologists can ordinarily differentiate fossil records at approximately 100,000-year intervals. Anything that occurs on the evolutionary timescale in a smaller time interval is usually beyond the level of detail that the data can confirm or disconfirm. This means the loss of enormous detail in the evolutionary sequence (recall how many species have become extinct in the last 100 years alone), yet we can still draw coarser inferences that are of great value based on the available level of resolution.

These constraints on the level of detail with which a problem can be treated are a result of inherent limitations on what can be achieved computationally. In modeling turbulent flow, realistic problems such as the flow of air over a wing require Reynolds numbers of the order 10^7. To simulate this digitally would require a grid with around 10^{23} points, a number comparable to the number of molecules in the fluid, and completely inaccessible on foreseeable computers.[24] The reason for this is that most of the energy in turbulent flow is carried in very small eddies that require extremely fine grids to simulate fully.[25] Although there currently exist eddy simulation models of clouds with a resolution of tens of meters, the enormous computing demands it would require to extend these simulations to the large regions required by climatological models makes the extension impractical.[26] A complete simulation for a major storm would

[23]Ibid., p. 7.

[24]As of 2003, Reynolds numbers of the order of 10^4 are computationally feasible. The Reynolds number is a dimensionless constant characterizing the flow. For a pipe, it is the diameter of the pipe multiplied by the average velocity of the fluid, divided by the viscosity.

[25]*Science* 269 (September 8 1995): 1361–1362; Tajima 1989.

[26]See Petersen 2000.

require the representation of phenomena ranging from a size of a few millimeters to regions of up to 100 miles, which would require more than 10^{16} times the resolving power of contemporary simulations.[27] These constraints on resolving power thus require that only phenomena appropriate to the available level of resolution be modeled, a contemporary example of our second principle from section 3.2.

I have argued that scientific progress in extending the realm of the observable depends to a considerable degree upon technological improvements. The refinement of existing instruments and the invention of new ones continually, and sometimes radically, extends our access to the material world. Perhaps less obvious is the idea that progress in theoretical science, in large areas of applied mathematics, and in restricted areas of pure mathematics now depends upon technology. The following figures are revealing in this regard.[28] Abstract developments of quantum mechanics require an infinite set of basis vectors to represent states. For the finite basis sets that actual applications need, suppose that m atomic orbitals are used (in the linear combination of atomic orbitals' representation of molecular orbitals—the LCAO method). Then one needs $p = m(m+1)/2$ distinct integrals to calculate one-electron Hamiltonians, and $q = p(p+1)/2$ distinct integrals to calculate electron interaction terms. This gives

$$m = \quad 4 \qquad 10 \qquad 20 \qquad 40$$

$$q = 55 \quad 1540 \quad 22155 \quad 336610.$$

With this number of integrals to be calculated, computational constraints, which primarily involve memory capacity, place severe limitations on what can be done to implement theory at any given stage of technological development. Such constraints are different in principle from the constraints that the older analytic methods put on model development, because for the other methods mathematical techniques had to be developed to allow more complex models, whereas in many cases in computational science, the mathematics stays the same and it is technology that has to develop.

In finite difference simulations, the degree of resolution is limited by the speed of the computational device and usually also by its memory capacity, and as these two features are improved, so is the quality of the simulation. Improvements in these areas often mean that systems

[27]See, e.g., Heppenheimer 1991.
[28]McWeeny and Sutcliffe 1969, p. 239.

previously simulated in two spatial dimensions can henceforth be simulated in three dimensions, or that the dimensionality of the phase space underlying the computational model can be significantly increased. There are analytic models of accretion disks in astrophysics, but for three-dimensional models, simulations are needed because disks are not spherically symmetrical in the way that stars generally are. A two-dimensional analysis of disks thus omits essential information that a three-dimensional simulation can provide, and the move from the one to the other can be made only when technological progress permits.[29] In much chemical modeling, there is also an essential three-dimensional element because the surface of the molecule does not give the appropriate information about its structure. In the early 1970s it was the two-dimensional topographical features of molecules that served as the focus of pharmaceutical research, but the limitations of this approach quickly became evident and three-dimensional structural features are now the main focus of activity in that area.[30] As our chapter-opening quote illustrates, the development of molecular modeling frequently has been crucially dependent upon hardware improvements.

Given that the use of models of dimensionality lower than that possessed by the real system is often forced by technological constraints, are those models then essentially defective? Not always, because certain qualitative features, such as phase transitions, can occur in the lower-dimensional models, and it is thus possible to demonstrate that those mechanisms can account for the phenomena in the smaller-dimensional model. This is an example of a particular kind of simplification that we can call dimensional scale reduction.

We can thus see, through the vehicle of the microscope metaphor, how the progress of theoretical science is dependent upon technological extensions of our mathematical abilities. The realm of the tractable, like the realm of the observable, is an expanding universe.

4.4 Observable, Detectable, and Tractable

In the case of material entities, the simple, fixed, and sharp division between observable entities and unobservable entities fails to capture

---

[29]Hawley 1995, p. 1365.

[30]"The second generation has shown that consideration of the full detailed properties in 3-D is necessary in allowing the subtle stereochemical features to be appreciated" (Cohen 1996, p. 3).

important differences in the nature of the subject matter and its relationship to us because the degree of technological difficulty involved in
moving things across the boundary is relevant to what is observable in
practice. Some humanly unobservable entities are quite easy to bring into
the realm of the instrumentally observable—eighth-magnitude stars, for
example—whereas others, such as neutrinos, are difficult in the extreme.
Some, such as ω particles, require considerable theoretical description to
bring them into contact with human observers, whereas others, such as
wood cells, do not. Certain sorts of 'unobservables' are merely theoretical
conveniences that, although they could be observed were they to exist, are
never seriously considered to be real. Finally, there are the difficult cases,
troubling for realists but hard to set aside. The nature of gravitation is
a case in hand. One could argue that gravitation is an observable—we can
detect it unaided through muscle tension, even when sight, hearing, taste,
smell, and touch have been neutralized—but notoriously, what its true
nature is has been subject to multiple redescriptions over the centuries.
Selective realism allows us the flexibility to reflect these degrees of
observational difficulty in different attitudes toward the reality of the
hypothesized entities or properties.

Once we move from human observability to instrumental detectability, and insist that what is important for science is what is detectable in
practice rather than what is detectable in principle, the idea of a sharp,
uniform, permanent boundary between the observable and the unobservable has to be abandoned. As we have argued, the boundary between what
is detectable in practice and what is undetectable in practice is not at all
fixed, but is a moving surface that is a function of technological progress.
Moreover, that moving surface is highly irregular, with some areas having
deep salients beyond the humanly observable, and others barely pro-
gressing past it. The 'detectable in principle' tag is, of course, important
for certain purposes. For what is undetectable in principle, undetectable
by any physically possible instrument, must be causally inert. That would
not necessarily put it outside the realm of science, for science routinely
uses terms that purport to refer to abstract objects, but for such things
a different kind of realism than the one suggested here is required.[31]

---

[31]Azzouni 1997 argues that four criteria for something to count as an observation carry
over to reliable instrumental detection. These criteria are (1) the access must be robust, in
the sense that it operates more or less independently of what we believe; (2) they can be
refined; (3) they enable us to track the object; and (4) certain properties of the object itself
play a role in how we come to know (possibly other) properties of the object.

There is an important difference between enlarging our native abilities in the ways we have described, and counterfactually supposing that we had those enhanced abilities from the start. If we were equipped with built-in electron microscope eyes,[32] the problem for us would be the inverse of what it now is; it would be the question of how to infer what the macroscopic world looks like from the images presented to us by the electron microscopes. This would be a tremendously difficult task, one that would require synthesizing the macroscopic world from the microscopic as well as predicting emergent properties, and it indicates how different the course of science would have been had we been so equipped. Moreover, we should need to invent devices that would get us from the direct perception of the molecular level to images of mountains, human beings, white-water rivers, and so on.

The situation is a little different when we compare computability in principle and computability in practice, for there are objects of legitimate mathematical interest that are uncomputable in principle, and hence the question of which functions are computable in the traditional 'in principle' sense is a legitimate one for both pure mathematics and philosophy. But once we turn to applied mathematics, considerations parallel to the case of detectability apply. When philosophers use the term 'computable', they almost always have in mind the purely theoretical sense centered around Church's Thesis. This thesis identifies the number-theoretic functions that are computable in some generally recognized but informal sense with those which are computable according to a set of precisely defined criteria, such as being a general recursive function, being Turing machine computable, and being computable in the sense of Post. It is provable that each of the precise criteria picks out the same set of functions, hence suggesting that each has captured, from a different perspective, the correct concept of 'computable'. But each of those precise criteria concerns 'in principle' computability, abstracting from real-world limitations such as bounded memory, speed of computation, and so on. The powerful results in recursion theory could not have been achieved without such abstraction, but the concerns of the logicians are different from ours. It is better to replace the term 'computable' with the term 'computationally tractable', construed as involving what can be computed in practice on existing computational devices.

---

[32]Paul Churchland's suggestion (1985) is rather different—he imagined a creature with one human eye and one electron microscope eye.

The situation with the concept of computational tractability is in some ways the inverse image of the situation regarding what counts as observable. Minimalist empiricists had a very narrow conception of what is observable, whereas the domain of what is considered here as observable with the aid of technological advances is much larger. With the concept of computational tractability the wide domain of recursive functions is drastically reduced in size to obtain the set of functions that are tractable with the aid of contemporary computational science. But within the domain of this technologically enhanced conception there remains a parallel, for what is considered computable goes far beyond what is actually calculable by the limited powers of a human, and it is not a permanent boundary but one that is a function of technological progress.

It should be noted that there is a double element of contingency in this appeal to actual computability. It is obviously contingent upon what is currently available to us in terms of artificial computational devices, but it is also contingent upon the complexity of the world itself. For, if the number of degrees of freedom involved in physical systems was always very small, we should not need augmentation of our native computational abilities to employ such procedures as Monte Carlo methods. But the world is not simple in that particular way, and in order to apply our theories to systems with a large number of degrees of freedom, such augmentation is forced upon us. This double element of contingency is not a reason to reject the position for which I have argued. In fact, it is exactly the kind of consideration that should place constraints on an empiricist's epistemology, because what is acceptable to an empiricist has to be influenced by the contingencies of our epistemic situation rather than by appeal to superhuman epistemic agents free from the constraints to which we, or anything remotely like us, are subject.

So, there is a significant parallel that can be drawn between the issues involved in instrumentational enhancement and those involved in computational science. The most immediate conclusion to draw is that in dealing with issues concerning the application of mathematical models to the world, as empiricists we should drop the orientation of an ideal agent who is completely free from practical computational constraints of any kind, but not restrict ourselves to a minimalist position where what is computable is always referred back to the computational competence of human agents. If, as many minimalist empiricists believe, it is impermissible to argue that humans could have had microscopes for eyes or could evolve into such creatures, and it is impermissible so to argue because empiricism must be concerned with the epistemological capacities of humans as they are presently constituted, then it ought to be equally unacceptable to argue that in principle humans could compute at rates 10^6 faster than they actually do.

4.5 Other Kinds of Simulation

Our primary focus has naturally been on simulations run on digital computers, but there are other kinds of simulations that play a role in science. I shall show that they all fall under our general conception of a computer simulation and that they are unified methodologically by their various uses of computational models in the sense of section 3.14.

Analog computers were dominant in the early work on computing machines in the 1940s, but declined in popularity in part because analog simulators are chiefly useful for linear ordinary differential equations with constant coefficients; nonlinear characteristics, which are relatively easy to deal with via digital simulations, often require expensive special-purpose hardware in analog simulators. In addition, analog computers tend to be special purpose, as opposed to the flexibility of digital computers, and they are limited in accuracy to about 0.01%.[33] In fact, it is rare to find problems that are solved equally well on both analog and digital machines—their virtues and vices tend to be complementary. Analog machines briefly reemerged in the popular consciousness because of their use in some of the original analyses of chaos theory and in distributed models of cognitive function, although the overwhelming emphasis in computational science is still on digital computation. Finally, there is some interesting work on mathematical characterizations of analog computability and its relations to discrete computability,[34] which forms an abstract counterpart to these concrete analog systems. Workers in the area sometimes make a distinction between analog computers and analog simulators, but this distinction is largely a result of how the machines are used, rather than of any intrinsic differences between them, and this difference will henceforth be ignored.

---

[33]See Ord-Smith and Stephenson 1975, pp. 135–139 for a further comparison of the two approaches.

[34]See, e.g., Rubel 1989; Earman 1986, chap. 6.

The essence of an analog computer is to set up a physical system whose behavior satisfies the same equations as those which need to be solved. By adjusting the input to this system, and measuring the output, the latter will give the solution to the desired equation(s). Recalling the fact mentioned in section 3.5 that a small number of equation types covers a vast number of applications, analog computers fit neatly into the picture suggested there of scientific organization. Figures 4.1 and 4.2 show one example of analog computation/simulation, based on a example from Karplus 1958, chapter 1.

On the left is a damped oscillator (figure 4.1); on the right is an electrical circuit that simulates the motion of the oscillator (figure 4.2). The mass M is associated with the inductance L, the damping force D with the resistance R, and the spring force K with the reciprocal of the capacitance C. The external force F on the mass is associated with the source voltage V, and $q(t)$, the charge, is given by

$$q(t) = \int_0^t I \, dt,$$

where I is the current. The equations governing the two systems are then

$$F(t) = Md^2x/dt^2 + Ddx/dt + Kx \tag{4.1}$$

$$V(t) = Ld^2q/dt^2 + Rdq/dt + q/C. \tag{4.2}$$

It can easily be seen that the form of the resulting equation is the same for both systems, even though Newton's Second Law is used in the

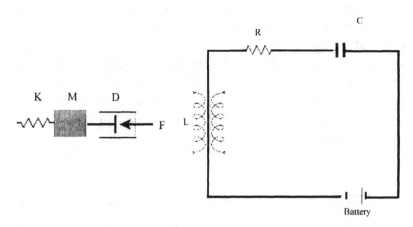

Figure 4.1   Figure 4.2

derivation of equation (4.1), whereas Kirchhoff's Law underlies the derivation of equation (4.2).[35] This identity of mathematical form allows the electrical circuit to simulate the operation of the damped oscillation of the spring, despite the gross physical dissimilarities between the components of the two systems. The core simulation here is the dynamic process by means of which the simulator 'solves the equations' in virtue of its physical behavior. There is also a very simple representational display of the damped oscillator's behavior by the oscillation of the voltmeter needle, resulting in a full, albeit crude, simulation of the damped oscillation.

Although in analog computers the physical operation of the computational device might seem more directly relevant to how the simulation is carried out than it is in the digital case, in both digital and analog simulations the relevant similarity occurs only at the level of the formal structure represented in the underlying mathematical model. If we want a unified theory of simulation that covers both analog and digital simulations, the theory must be at the level of the unifying computational template and not at the level of implementation.[36]

Modern (i.e., post–World War II) analog computers almost always use electrical circuits as the physical system. There are two advantages to this. One is practical, the speed of electrical analog devices. The second is the analytic aspect of circuits—the ability to break down the behavior of the system into basic components, the behavior of which is known, and to then construct the required system from these basic components. Thus, one avoids the question "How do you know that the computer actually does satisfy the equations?" in virtue of applying our dictum from section 2.8 to know how the instruments work. In general, this would be a tricky question, because in cases where one cannot solve the equation analytically, comparison of the computer's behavior with the equations would be as difficult as a comparison of the original system's behavior

---

[35]"The analogy is rooted in the fact that Newton's Law ... and Kirchoff's [*sic*] Law ... represent applications of the same basic principles that underlie most areas of physics—the laws of energy and continuity. It is the translation of these laws into mathematical language that leads to the characteristic differential equations and since the same laws are applicable to the mechanical and electrical systems, the characteristic equations are similar in form" (Karplus 1958, p. 11).

[36]Because simulators are based on (formal) analogies, the position taken by Hesse 1966, p. 20—"What ripples and sound waves have in common is completely contained in the mathematical formalism, and it is this we point to by continuing to use the word 'wave'"— has close affinities here.

with the equations, and nothing would have been gained. But it is knowing what constitutes the structure underlying the simulator, together with a small set of fundamental laws of electrical circuits, that comprises the construction assumptions for the simulation template that allows us to have confidence in accuracy of the analog simulation. An added bonus here is the flexibility that this allows in applying the same basic components to simulate different systems. This modularity of analog simulators means that the justification of the construction assumptions behind the simulation reduces to having a justification for the mathematical representations of the basic units and of the principles of combination. In particular, the ability to reduce nth-order ODEs to systems of first-order ODEs plays an important role in the process.[37]

Here I note just a few points that are specific to analog simulations. The behavior of continuous systems is modeled exactly by analog simulations, unlike digital simulations, within which there are always discrete approximations to continuous functions. The approximations in analog computation arise because of physical limitations on the components in the simulator, such as noise, imprecision of measurement, and so on. This means that graphical representations are preferable to data displays, because the precision of graphical representations that can be assessed with the naked eye approximates the real precision of the simulation.

A different kind of simulation, scale simulations, uses a concrete model of a different temporal or spatial size than the system that is being simulated. The archetypal case here is the use of wind tunnel experiments on scale models of aircraft and cars. Ordinarily, the simulation models are smaller than the real thing, as when laboratory gas jets are used to simulate astrophysical jets or when *Drosophila* are used to study genetic changes over a faster timescale than occurs in other species, but they also can be larger, as when, imperfectly, the diffraction of light waves is simulated by fluid waves. The crucial requirement for scale simulations is the scale invariance of their behavior, and this scale invariance can once again be represented by the fact that the same computational templates cover the simulating and the simulated systems, with parameter values differing between the two.

Despite the interest of analog and scale simulations, because most contemporary computer simulations are implemented on digital machines,

---

[37]Ord-Smith and Stephenson 1975, p. 13.

I shall henceforth restrict myself to those.[38] Thus, from here on, by "computer simulation" I mean a simulation that is implemented on a digital computer.

4.6 Agent-Based Modeling

One of the most serious problems facing mathematical modelers in the social sciences, specifically sociology, economics, and anthropology, is the complexity and transience of the systems being modeled. These issues, which are also prevalent in population biology and ecology, can be at least partially addressed through the use of computer-assisted modeling and simulation. But before computationally modeling such systems can begin, a decision must be made, whether explicitly or implicitly, about whether or not to adopt methodological individualism. What counts as methodological individualism is notoriously vague, but it standardly involves the position that all social relations between individuals must either be reinterpreted in terms of intrinsic properties of those individuals, together with permissible physical relations such as spatial relationships, or, in more recent treatments, that those social relations must be shown to supervene upon intrinsic properties of individuals. That is, on the individualist view, it is impossible to have two groups of N individuals which are indistinguishable with respect to their intrinsic properties but which have different social relations holding between their members.[39] As a result, the existence of autonomous social facts is denied in such individualistic approaches. Since there are no social facts on this view, there is no causal influence from the group or social level to the individual level. In contrast, models that allow for real social relations—or as I prefer, models within which emergent social phenomena appear—usually include some kind of downward causation from a social structure to the individuals operating within that structure.

[38]A more controversial use is the employment of animal models in biomedical research to simulate human physiological processes in calculating, for example, dose/response ratios for mutagenic substances. The choice here is usually made on grounds of similarity of physiological, cellular, or genetic structure, rather than on the basis of the identity of mathematical form. For a criticism of such models, see Freedman and Zeisel 1988. I shall not discuss here the often emotional debates concerning the extent to which animal models can be replaced with computer simulations.

[39]In contrast, the spatial relations between identical pairs of individuals obviously can be different. In this sense, spatial relations are external. There is a close connection between denying social facts and asserting that all social relations are internal.

An increasingly important type of contemporary computer simulation is explicitly built upon such individualism in the form of agent-based or individual-based models, which are descendants of the cellular automata models developed by von Neumann and Ulam.[40] Models developed along these lines can be found in Thomas Schelling's early work on segregation,[41] in the more recent treatments of Prisoner's Dilemma situations by Robert Axelrod,[42] in the Swarm modeling programs,[43] and in many other applications.[44] Agent-based models are by no means restricted to the social sciences. Other applications of cellular automata involve lattice gas automata to model ferromagnetism and simulated annealing; modeling autocatalytic chemical reactions; modeling the flocking behavior of birds, fish, and other species subject to predators and for which local groupings can lead to aggregate formations that are optimal for the flock as a whole; investigating termite behavior in arranging wood debris; and modeling business alliances for promoting technical standards.[45] It is a central feature of such models that only local influences on individuals are allowed, these influences being governed by simple rules.

Agent-based models can overcome one of the primary difficulties that affect traditional template-based modeling methods—the fact that the agents are operating within an environment which is constantly changing, and the fact that an agent's actions are reciprocally affected by the choices made by other agents. In addition, in complex adaptive systems, agents can be subject to learning or mutation and selection. It is here that we can see the significant advantages of the specifically dynamic aspect of computer simulations discussed in section 4.1, because agent-based models iteratively compute the state of the system at each time step in the computation and thus can accommodate such dynamic changes. At the very least, such possibilities greatly expand the number of available models. Most traditional modeling methods have to consider the population as homogeneous; each individual in the population is taken to have the same basic properties as any other. This constraint need not be imposed

---

[40] John Horton Conway's game Life is also a direct descendant of cellular automata models.

[41] Schelling 1978.

[42] Axelrod 1997.

[43] For references see http://www.swarm.org or http://www.santafe.edu.

[44] See, for example, Kohler and Gumerman 1999; Wolfram 2002. Some of the claims made in the latter work are controversial.

[45] See Wolfram 1986; Axelrod 1997.

upon the more flexible, individualistic, agent-based models. Within modern societies, agents are often quite diverse, and the constraints and influences on the social systems are changing quite rapidly. One such example involves business models where computer companies are the agents. The business environment in such an industry is fluid and highly dynamic, with changing expectations on the part of consumers, evolving standards, the constant appearance and disappearance of competitors, internal changes in the companies themselves, and so on.

In a more radical way, agent-based simulations can lead to a fundamental reorientation of a field of inquiry. For example, in Epstein and Axtell's Sugarscape models[46] it is possible to model changing preference orderings for the agents, leading to a significant degree of disequilibrium in the system compared with the equilibria that generally result from the fixed preferences of traditional economic models.[47] Effectively modeling such systems requires a resolutely dynamic approach that cannot be captured using traditional static mathematical techniques.

This use of computers in modeling social systems is, however, distinctively different from their use in the kinds of computer simulations that we have considered up to this point. In those kinds of simulations, an explicit mathematical model is constructed before the simulation begins, and the computational power is brought to bear to produce solutions for the equations that form part of the underlying model. Within agent-based models, in contrast, there is no overarching model of the social system, only rules governing the interaction of the individuals comprising the system. We have here a variant of the types of model we discussed in section 3.10, in which little or no systematic theory governs the construction of the underlying model. Phenomena often emerge within these models at the macro level that are the result of multiple interactions at the micro level, these macro phenomena being unforeseen when the individual-level analysis is begun. It is the appearance of these higher-level phenomena that makes methodological individualism a relatively uninteresting epistemological position. For it is often only in virtue of examining and conceptually representing a system above the level of the individuals that novel information about the system becomes perceptible. And one can mean 'perceptible' literally, for visualizations based on the output from the simulation are often essential in discerning these macro

---

[46]Epstein and Axtell 1996.
[47]For other examples in the same area, see Skyrms 1996.

patterns, another example of the issue concerning simulation outputs that we discussed in section 4.1. Ontologically appealing as individualism may be, it can be epistemically misleading about the kinds of information we can extract from the simulation.

One of the more important questions that arise about agent-based modeling is the degree of understanding which is produced by the models. It is often claimed that rather than giving a realistic account of social phenomena, what these agent-based models can do is to provide us with clarification of, and insight into, the social phenomena they model, and that the primary goal of such simulations is to reproduce broad qualitative features of the system from a few simple assumptions about the agents. As Epstein and Axtell put it: "As social scientists we are presented with 'already emerged' collective phenomena, and we seek micro rules that can generate them."[48] This exploratory approach is not unique to agent-based models. Adjustable-parameter models, such as those which are used in accretion disk models in astrophysics, allow the parameters and functional forms to be adjusted to fit the data, although this renders them immune from falsification.[49] Another example is the study of solar circulation,[50] where the simulations are used to develop physical intuitions about the solar circulation, rather than a detailed description.

In fact, as we shall see in the next section, because the goal of many agent-based procedures is to find a set of conditions that is sufficient to reproduce the behavior, rather than to isolate conditions which are necessary to achieve that result, a misplaced sense of understanding is always a danger. Agent-based models are a powerful addition to the armory of social scientists, but as with any black-box computational procedures, the illusion of understanding is all too easy to generate. Similar sentiments have been voiced about simulations of turbulent flow that provide no real insight into the underlying physics.[51] These instrumental models are best viewed as examples of constraint satisfaction modeling, within which a wide variety of behavior can be exhibited, provided it conforms to a set of quite general constraints—in the example we have just cited, these constraints will be the rules governing the individuals and, in the main example of the next section, the restriction that all orbits

---

[48]Epstein and Axtell 1996, p. 20.
[49]Hawley 1987, p. 677.
[50]Brummell et al. 1995.
[51]See Heppenheimer 1991, p. 39.

must be compounds of uniform circular motions. The constraints may be natural, as in physical systems, or artificial, as in simulations.

The moral to be drawn here is that the use of simple mechanisms is acceptable for purposes of understanding only if there is good reason to hold that these "mechanisms" correspond to genuine features of the systems under investigation. Certainly, quite sophisticated physical models employ radically simplified physical mechanisms to construct descriptively accurate computational models. One example is that of Ising models in physics to describe ferromagnetism (see section 5.3). In these models, a common simplification is to restrict the interactions between particles on a lattice to nearest neighbor interactions and to reduce the continuously variable orientation of the spins of those particles to binary 'up' and 'down' values. Despite these severe simplifications, indeed oversimplifications, such models have been quite successful at reproducing major features of ferromagnetism and other phenomena in solid-state physics. But it would be a mistake to move from accepting the partial predictive effectiveness of these models to thinking that they provide us with an understanding of the underlying systems. The power and flexibility of agent-based models can thus render them subject to as much abuse as more traditional forms of modeling.

4.7 Deficiencies of Simulations

> I sometimes have a nightmare about Kepler. Suppose a few of us were transported back in time to the year 1600, and were invited by the Emperor Rudolph II to set up an Imperial Department of Statistics in the court at Prague. Despairing of those circular orbits, Kepler enrolls in our department. We teach him the general linear model, least squares, dummy variables, everything. He goes back to work, fits the best circular orbit for Mars by least squares, puts in a dummy variable for the exceptional observation—and publishes. And that's the end, right there in Prague at the beginning of the 17th century.[52]

We have described many of the advantages of computer simulations. What, in contrast, are their dangers? There is the danger of a return to a priori science. The very cheapness and ease of simulations, compared with laboratory or field experiments, tempt one to replace the latter with

[52]Freedman 1985, p. 359.

the former, and there is some feeling in the experimental physics community, for example, that these practical considerations are leading to a preferential use of simulations even in situations where they are inappropriate. Tuning the parameters in a simulation in order to fit preexisting data is a constant temptation with these methods. We can see how dangerous this is by using an example similar to the one described in the quotation above.

As we have seen, it has been claimed for agent-based models that one of their primary uses is exploratory, in the sense that it is of interest to show that simple rules can reproduce complex behavior. But this cannot be good advice without imposing extra conditions. Consider a small thought experiment. Kepler's planetary model of simple, continuous, elliptical orbits replaced the Ptolemaic and Copernican models that both used the mathematical apparatus of epicycles and deferents. How and why this happened is an interesting and complex story, but as is reasonably well known, there was no inevitability about the transition because in principle, the apparatus of epicycles and deferents had the mathematical power to capture any observed data about the motions of the planets. More precisely, if the motion of a planet can be represented by a holomorphic function $f(t)$ (i.e., f is single valued, continuous, and differentiable on some annulus $0 \leq R_1 < |z - a| < R_2 < \infty$ so that f is of bounded variation), then for any ε, there is some superposition of epicycles $S_M = \sum_{j=1}^{M} c_j \, e^{i a w_j t}$ such that $|f(t) - S_N(t)| < \varepsilon$ for all $N > M$. Now suppose that early seventeenth-century astronomers had been given access to modern computer simulations and Kepler, being the smart guy that he was, had enthusiastically bought into those techniques. Then Kepler would have saved himself ten years of heartbreaking labor and been able to easily account for Tycho's data with a bit of artful parameter tweaking. And that's the obvious danger with the exploratory use of any such models. Because it is often possible to recapture observed structural patterns by using simple models that have nothing to do with the underlying reality, any inference from a successful representation of the observed structure to the underlying mechanisms is fraught with danger and can potentially lock us into a model that is, below the level of data, quite false.

In a different direction, because the simulations are entirely under the control of the simulator, the assumptions of the underlying model are, barring incompetence, built into the simulation and so the simulation is guaranteed to reproduce the results that the model says it should. In some cases, this is unproblematical, but when simulated data rather than real data are fed into the simulation, the prospects for informing us about the world are minimal. To take a simple example, consider a simulation of tossing a coin. Here the independence of the tosses and the equal probabilities of the outcomes are built into the output of the pseudo-random number generator (which here would be modified to produce only the outputs 0 and 1, representing heads and tails, respectively), and so it would be pointless to use such results as a test of whether the binomial distribution accurately modeled real coin tosses. While in this case the pitfalls are obvious, there are other, more subtle, cases in which statistical simulations will do nothing but explore the properties of the random number generator that is being used.

Such cautions are of more than academic interest. The Millennium Bridge in London that connects the north and south banks of the Thames is a suspension bridge of innovative design that can move laterally. It has long been known that standing waves can build up in suspension bridges from vibrations; the Albert Bridge in central London to this day has a sign ordering soldiers to break step when crossing the bridge. Even though the Millennium Bridge design had undergone extensive computer simulations, it had to be closed shortly after its official opening when swaying from the unexpectedly large number of pedestrians became dangerous. The chief engineer at the architectural firm that designed the bridge ruefully concluded, "The real world produced a greater response than we had predicted. No one has effectively studied the effect of people on a horizontally moving surface."[53]

[53]http://www.telegraph.co.uk/news/main.jhtml?xml=news/2000/06/29/nbrid29.xml

5

Further Issues

5.1 Computer Simulations as Neo-Pythagoreanism

The feature that makes templates such a boon to scientists, their extreme flexibility of application, makes them a bane to scientific realists. The reason for this is straightforward. A great deal of attention has been paid by philosophers to the problem of *empirical underdetermination*, the problem that for any set of empirical evidence, there are many different, incompatible, theories, each of which is supported by the evidence. This is a problem of great generality, for, as Barry Stroud pointed out,[1] many traditional problems of skepticism result from empirical underdetermination. Take the entirety of your sensory experience. One theory, the commonsense view, says that this is the result of events occurring in an external world as we ordinarily conceive of it. A rival theory suggests that these experiences are brought about by a sophisticated virtual reality apparatus controlled by a religious order the main purpose of which is to provide the appropriate temptations for testing freedom of choice. The virtual reality apparatus is so good that no amount of empirical data will enable you to see through these appearances to the real world, which consists mostly of dense fog, an utterly bland place within which there are no real temptations, only the appearance of such. That is an entirely fanciful scenario, but more serious cases are of direct relevance to scientific practice.

Situations involving empirical underdetermination have two components that we can call *formal underdetermination* and *semantic underdetermination*. Consider a sequence of observations of the position of Mars.

[1]Stroud 1984.

Formal underdetermination consists in the fact that any given set of empirical data can be captured by multiple representations with different formal features while remaining consistent with that data. In the case of Mars, if we simply consider plotting the spatial and temporal positions of the observations, any hypothesis about the orbital shape that posits continuous variables will have discrete-valued rivals which cannot be empirically distinguished from the continuous model. More generally, for any function that is fitted to a set of data, there will be alternative functions that vary on values not yet observed, where 'observed' can be taken in whatever degree of liberality you wish. So, for any functional form that can be fitted to the available data, such as an ellipse, there exist other, arbitrarily complex, functions that differ from the elliptical representation at unobserved values. Of course, some of these alternatives can be eliminated by further observations, but there will never exist a unique formal representation. Note that the problem of formal underdetermination is independent of the nature of the subject matter; the fact that the required function represents the orbit of Mars, or indeed of a planet, is irrelevant.

Semantic underdetermination involves the other component of empirical underdetermination. Whereas in formal underdetermination there is a one–many relationship between a set of facts and formal representations of those facts, in semantic underdetermination there is a one–many relationship between a given formal representation and different sets of facts.[2] When a theory is cast into a purely formal mode, it is always possible to find an interpretation for that theory which makes all the statements of that theory true, including those which describe the data, but which is different from the interpretation intended by the deviser of the theory and which is attached to the original representation. That is, the formal theory can be reinterpreted, often radically, so as to be about a different subject matter, yet remain true.

This problem of semantic underdetermination is not unique to the syntactic account of theories, for a similar problem affects the semantic account, and it provides perhaps the most serious difficulty for any realist interpretation of theories. It is also impossible to formally capture the intrinsic nature of the subject matter in these formal approaches, because the emphasis is on capturing the abstract structural relations behind the theory. The slogan 'Set-theoretical models are determined only up to isomorphism' encapsulates this semantic underdetermination problem,

---

[2]The situation applies even when the 'facts' are erroneous.

and the models are usually not even constrained that tightly because most theories are noncategorical (i.e., even the cardinality of the domain is not uniquely determined). Two models of a theory are isomorphic if there is a one-to-one, structure-preserving map from one model to the other. Informally, this means that the models have the same abstract structure and the intrinsic nature of the objects they are 'about' is not relevant to why the theory is true. A theory is categorical if any two models of the theory are isomorphic. Because of the limitations of many first-order formal languages, any theory with an infinite model will have models with larger (and/or smaller) infinite domains, thus precluding the one-to-one feature of the desired map.[3] This kind of semantic reinterpretation is what astronomers have done in a modest way with Kepler's original account of the planetary orbits. His own theory, which concerned magnetic celestial bodies, has been reinterpreted using our modern conception of the planets.

There are at least two versions of the semantic underdetermination argument, the radical and the moderate. The radical version proceeds by taking a given theory with its intended interpretation, stripping off that interpretation so that the theory becomes purely formal, and then showing that there are other, unintended, interpretations which make the formal theory true. The moderate version allows the interpretation of the observational terms to have a privileged status and to remain fixed, but this still leaves open multiple different interpretations of the theoretical terms, all of which are consistent with maintaining the truth of the observational claims.

Perhaps few people subscribe to the radical version of semantic underdetermination, and the moderate version tends to be a more pressing concern for scientific practice. Nevertheless, the specter of the radical view lives within any simulation. It was realized quite early in the twentieth century by mathematical logicians that standard formal languages could not uniquely determine a formal extensional semantics for a first-order theory, and it is well known that any consistent first-order theory can be given an interpretation the domain of which is either the set of natural numbers or a finite subset of them, depending upon whether the theory has infinite or only finite models.

---

[3]For a clear account of these concepts, see Mendelson 1987, and for an anti-realist exploitation, see Putnam 1980, a result that had already been noted by Max Newman in 1928—see Friedman and Demopoulos 1989.

Pythagoras and his followers believed that the world was made up of numbers,[4] and contemporary anti-realists have relentlessly pointed out that any consistent theory can be reinterpreted to be 'about' natural numbers. We have already seen the reason for this: Formal theories, when interpreted by extensional semantics, cannot capture the intrinsic nature of the properties represented by predicates, because all that such extensional semantics requires is that the elements of the domain be distinguishable from one another. If we call this kind of numerical reinterpretation *neo-Pythagoreanism*, we can see that computer simulations represent neo-Pythagoreanism in a particularly pure form. When presented with a formal theory, a digital computer operating on strings of binary states is simply interpreting the theory in purely numerical terms. The strings could be reconstrued as many other things as well, because calling the binary units "numbers" is merely a conventional courtesy. They are, in fact, nothing more than distinguishable strings of syntax. In this way, Pythagoreanism has been recaptured, dynamically, almost two and a half millennia after it was first proposed. This is precisely why Feynman's slogan of section 3.5 is the key methodological doctrine of computer simulations, and why any attribution of realism to simulations has to be in terms of the full computational model underlying the simulation because the model contains a nondetachable interpretation (see section 3.14). The realist commitments are generated by the construction process—they can never be read from the simulation itself, however vivid the graphics.

It is a serious mistake to consider the simulation as lacking whatever interpretation was originally employed in constructing the template. To advocate semantic amnesia, as have many modern philosophers, is to deny that the historical development of a theory, with particular interpretations attached to the formalism, has any relevance to our acceptance of the theory. The semantic amnesia view treats theories as ahistorical objects that can be examined at any given point in time without regard to their development. In contrast, as we urged in section 3.7, it is how these computational models are constructed that is crucial to why we accept them and to what they pick out.

This position, as it happens, is not at odds with the views of the very logician upon whose model-theoretic work some of these anti-realist

---

[4]Geometrical analogies with square numbers, triangular numbers, and so on made this rarefied hypothesis moderately credible.

claims have been founded. In his article "The Concept of Truth in Formalized Languages," Alfred Tarski insisted:

> It remains perhaps to add that we are not interested here in "formal" languages and sciences in one special sense of the word "formal", namely sciences to the signs and expressions of which no material sense is attached. For such sciences the problem here discussed has no relevance, it is not even meaningful.[5]

5.2 Abstraction and Idealization

The fact that instruments are property detectors (section 2.7) sheds some light on certain traditional philosophical issues. All computational templates require us to abstract from the systems that the templates represent, and most require us to make idealizations and approximations. Saying what this means requires us to be clear on what are idealizations and what constitutes the process of abstraction—or, as I shall from now on simply call it, abstraction. (I shall have little to say here about approximations and nothing about simplification, for in modeling contexts 'simplification' tends to be merely a loose term covering the uses to which abstraction and idealization are put.)

The term 'abstraction' has been used in a variety of senses; here I shall identify four. Three of them are covered by a process that we can call property elimination, which takes us from whatever total collection of properties[6] is possessed by an entity to a proper subcollection of those properties, those properties being left unchanged.[7] Abstraction thus requires properties to be modular in the sense that the remaining properties are invariant when any given property is eliminated. Property elimination can be carried out in three different ways. In the traditional empiricist sense of abstraction, the process of property elimination is a cognitive

---

[5]Tarski 1956, p. 166. Even Nelson Goodman, well known for promoting radical underdetermination theses and an elegant developer of the themes canvassed in Carnap's *Aufbau*, insisted, as Geoffrey Hellman notes, that the primitives of a theory must have an interpretation: "A constructional system is understood to be an interpreted formal system.... The definitions of a constructional system are to be thought of as 'real' definitions meeting some definite semantic criterion of accuracy in addition to the usual syntactic criteria (eliminability and noncreativity) imposed on purely formal or "nominal" definitions.... The primitives are to be thought of as already having an intended use or interpretation." Goodman 1977, p. xx.

[6]'Properties' here includes relations.

[7]Except, of course, for the relation of the remaining properties to the now absent ones.

operation that involves focusing attention on one or more properties of
a concrete object and eliminating from consideration all of the other
properties possessed by that object. In this sense, the remaining properties
continue to exist but are ignored. As we saw in section 2.7, this kind of
abstraction is carried out automatically by instruments, and our own
sensory apparatus is itself an abstraction device in that sense. Traditional
empiricism is thus, in virtue of this fact, committed by its very nature to
processes of abstraction — when we see a piece of radioactive uranium, we
consider it, unthinkingly, without many of the unobservable properties it
has, such as emitting α-particles.

One can, in a second sense, conceptualize the 'omitted' factors as
continuing to exist but having a value of zero, as is done when we reduce
to zero the Coulomb force between electrons in a model of the atom.[8]
Practically, the difference between the first and the second approaches
may seem to make no difference, but the difference of emphasis is of great
importance. In the first approach, omitting a property entirely suggests
that the modeler considers the property irrelevant. In the second approach, retaining the property while setting it to zero suggests that the
correction set will later adjust the value away from zero. These cognitive
processes of property elimination, although of great importance in the
development of models, have been replaced within formal representations
by a process of predicate elimination. In this linguistic variant of the first
two approaches, a given representation, usually formulated mathematically, will contain variables and parameters corresponding to only that
subset of properties considered relevant for the theory. It is this use that
underlies the kind of abstraction found in simulation models.

The third method of property elimination uses controlled experimentation to physically eliminate the effects of all properties except those
in which we are specifically interested. Unlike the first two approaches,
this method of abstraction intervenes in the world to physically remove
the abstracted effect. The dividing line between passively excluding properties because the apparatus is incapable of detecting them and actively
blocking them from affecting the apparatus may be a division with no real
basis. Because we usually distinguish between instruments and experimental apparatus, however, no harm is done in maintaining that separation here. There is a fourth use, involving randomized trials, that results

---

[8]Although note that some authors call this specific procedure, somewhat misleadingly,
"the central field approximation."

not in eliminating the property but in eliminating its causal effects on average by the use of random assignment of treatments. Because of the additional complexities involved in that case, I shall not discuss it further here.

The following definition of abstraction captures all three of the methods we want to consider:

> Abstraction₁ is the process of omitting properties or their related predicates from a system or a model, the omission process being either cognitive, representational, or physical, leaving the remaining properties or predicates unchanged.

An apparently different process sometimes goes under the name 'abstraction'. It involves moving from a specific instance of a property, or a structure, to the property or structure itself, considered as a universal. For example, Robinson claims, "Abstraction is becoming aware for the first time of a new general element in one's experience and giving it a name. A general element is a form or a pattern or characteristic as opposed to a particular or individual."[9] Goodman puts it this way: "The problem of interpreting qualitative terms in a particularistic system, of constructing repeatable, 'universal', 'abstract' qualities from concrete particulars, I call the problem of abstraction."[10]

To conform to this different use of the term 'abstraction', it appears that we need a second definition: Abstraction₂ is the process of moving from a specific instance of a property to the property itself. But there is no need for this second definition. Because we have identified individuals with collections of properties, abstraction₂ is just a special case of abstraction₁. It is what results when a process of abstraction₁ results in removing all but one of the properties constituting the concrete entity. Henceforth, I shall simply use the term "abstraction" for either process.

Abstraction is important because there is a prevalent view, descended from Leibniz and employed in a simplified form in Carnap's state descriptions, that requires every concrete object to possess a complete set of properties, both positive and negative. Within the Leibniz-Carnap view, removing even one property by abstraction will take the object out of the category of concrete, property-complete objects. But this view rests on two fallacious positions. The first is that the absence of a property P entails the

---

[9]Robinson 1968, p. 170.
[10]Goodman 1977, p. 106.

presence of its negation, not-*P*. The second is that *P* is a property if and only if not-*P* is a property. Unfortunate consequences have been drawn from these two positions. One is the view that the number of properties, positive and negative, associated with a concrete individual is so large that no concrete individual can be described completely. This in turn is taken to support the view that all scientific generalizations involve ceteris paribus conditions and that all linguistic descriptions must involve a process of abstraction.

Each of these positions is incorrect.[11] Consider an electron, which is a concrete entity. It consists in very few properties—charge, mass, spin, a probability distribution of position and one of momentum, perhaps a few others. Once we have specified these, we can truly say "It has these properties and no more." We do not need to add that it contains no fat, does not drink tea, and has never looked at an issue of *Sports Illustrated*. It does not have, and nomologically could not have, the property of being paranoid to any degree at all, even zero. Nomologically, because to be paranoid requires having mistaken beliefs, and electrons are nomologically incapable of having beliefs in virtue of their inner simplicity. Neither does an electron have the property of being not-paranoid. All that the phrase "and no more" entails is that the electron lacks the property of being paranoid. Anything else amounts to a reification of an absence, a nothing, a reification that constitutes a metaphysically unnecessary move. Perhaps we need an infinite set of basis vectors to fully describe the electron's state, but that is a peculiarity of the quantum specification and is not entailed by the concrete status of the electron. To insist that we must specify an infinite list of negative properties is to fall victim to a metaphysical position that has no place in science—or in ordinary life, for that matter.

Such claims are often resisted. Perhaps we have simply given up the logical neatness of property closure in exchange for a tidy definition of abstraction. Yet from the perspective of the realist, small-set concreta seem to be all around us, and physicalists in particular should see the force of the current view. For suppose that physicalism is true, at least to the extent that specifically mental properties do not exist. On the basis of that position, there is no such property, *S*, as believing that women's soccer is played more elegantly than men's. But if, on physicalist grounds, we reject *S*, why accept not-*S* as a genuine physicalistically acceptable property?

[11]Arguments against negative properties can be found in Armstrong 1978 and in Armstrong 1997, pp. 26–28.

And having rejected not-S for good physicalist reasons, it would be wrong to insist, simply on the grounds of logical closure or of property completeness, that any concrete particular such as a π-meson has the property not-S. The completeness feature has sometimes seemed necessary for describing nonactual situations in terms of sets of maximally consistent predicates, but even there it is unnecessary. We can say of a nonactual situation that a spatial region contains a solid red cube of dimension 2 cm on each side *and nothing else*. Nothing more needs to be said.

One joint consequence of property incompleteness and the property cluster account of objects adopted here is that abstraction applied to a concrete object does not necessarily produce an abstract object. By the term 'concrete object', I mean an object that has a spatial location, even if this spatial position is given only probabilistically,[12] and by 'abstract object' I mean one that does not have a spatial position. This allows some models to be concrete objects. One type of example occurs when a concrete object consisting of a smaller or different collection of properties serves as a model for another concrete object—an orrery is a model of the solar system; ball-and-stick models can represent molecules; the pig's cardiovascular system can model the human's. A model can even be an instance of the kind of thing it is modeling, because a model airplane can itself be an airplane, and an animal model for the human immune system itself has an immune system. The way in which models involve abstraction is simply in their possessing a subset of the modeled system's properties, even though concrete models will generally possess other properties that are absent in the modeled system. The process of abstraction can thus take us from a concrete object to a concrete object, from a concrete object to an abstract object, from an abstract object to another abstract object, but not from an abstract object to a concrete object.

There is another type of abstraction that is worth mentioning explicitly. Here, instead of omitting some of the properties, we arrive at more general versions of those properties. This occurs when we treat a burnt sienna object as being brown, or a planetary orbit is treated as a conic section rather than as an ellipse. This sense is closer to what people usually mean when they talk of the more abstract sciences. The relation between the more basic and the more general properties can be represented as a determinate–determinable relation, where A is a determinable

---

[12]I believe that this is sufficient; mental events, if they exist, are not concrete because they are not spatially located, even though they do have a temporal location.

of the determinate B just in case necessarily, whenever an individual has B, it has A.

These issues are of scientific as well as of philosophical interest. All planets travel in elliptical orbits and all comets travel in elliptical, hyperbolic, or parabolic orbits. Newton's genius united their motions under a general law: All objects subject to a central inverse-square attractive force travel in conic sections.[13] The general law does many things besides unifying those two facts. It greatly extends the domain of the laws from the motions of celestial objects near our sun to any object subject to gravity, and it also extends its application to nongravitational forces such as electrostatic or magnetic forces. In addition it turns descriptive, kinematical laws into a blueprint for dynamical laws, thus allowing causal explanations. It does all these things by using the more general category of conic sections to cover the more restricted categories of ellipses, hyperbolas, and parabolas used in the original laws, and also by finding something that planets and comets have in common, which is being acted upon by an inverse-square law. The first of these processes involves geometrical abstraction, which here takes the disjunctive category of elliptical-or-hyperbolic-or-parabolic and represents it as a determinable category, that of being a conic section, whereby disparate mathematical forms are unified by a more general mathematical form containing parameters, special cases of which give the original forms.

Why perform abstraction? Because in doing so, and focusing on properties, we move from a particular system with a large set of properties that may make it unique to a system with a smaller set that does not. Properties being universals, we are thus allowing the repeatability that is the essence of scientific method. The model represents not just the original system, but any system with those abstracted characteristics.

Turning now to idealization, we find that it is a different process from abstraction: Idealization involves taking a property and transforming that property into one that is related to the original but possesses desirable features which are lacking in the original.

---

[13]See Isaac Newton, *De motu corporum in gyrum*, discussed in Westfall 1980, p. 404. In that work Newton proves that an inverse-square law entails a conic orbit. But he did not, apparently, do this by abstraction, nor to deliberately unify the laws (the generalization about comets was not known until after Newton's work). It was simply a demonstration that the orbital shape is a consequence of assuming the inverse-square law. So although abstraction could allow us to state the law in an elegant way, in this case it was not necessary either for its discovery or for its statement.

The term 'desirable' indicates that idealization is not an objective process—what may be an idealization from the perspective of one set of goals may not be one from the perspective of another. Examples of idealization are treating planets as perfectly spherical rather than as spheroidal, treating collections of discrete objects as continuous, treating noninstantaneous collisions as instantaneous, and so on. In all of these cases, the idealization is applied to a property, and only indirectly to an object, but because the property remains as a constituent of that object, idealization can be applied without abstraction. It is obvious that, in the converse direction, abstraction can take place without idealization. Although it is possible to speak of whole systems as ideal, as when a die has been carefully machined to a symmetry that is lacking in mass-produced dice, the term 'ideal' is usually used in a sense that represents a physically unobtainable end state. Idealizations thus usually involve models of systems, and I shall so restrict my use of the term here.

We should not require that idealization always involves simplification. Simplicity is a notoriously unclear concept, and although idealization often results in simplification, it can also result in a more complex property than the one with which the process began. For example, in Bayesian theories of subjective probability and utility it is often assumed, as part of the concept of an "ideally rational agent," that the agent has a degree of belief about any proposition whatsoever. Such an agent is far more complex than are real humans, who often do not have any views at all about a given proposition. Some authors have not made clear the difference between abstraction and simplification: "Abstraction consists in replacing the part of the universe under consideration by a model of similar but simpler structure."[14] Although this covers some cases of simplification by abstraction, it is misleading, because one can achieve simpler models by retaining all the characteristics of a system but idealizing some of them, and by virtue of retaining all the properties, no abstraction will have taken place.[15]

5.3 Epistemic Opacity

In many computer simulations, the dynamic relationship between the initial and final states of the core simulation is epistemically opaque

[14]Rosenbluth and Weiner 1945.

[15]This claim relies upon some systems having only a finite number of properties. If we reject the closure of properties under arbitrary logical and mathematical operations, this is plausible.

because most steps in the process are not open to direct inspection and verification. This opacity can result in a loss of understanding because in most traditional static models our understanding is based upon the ability to decompose the process between model inputs and outputs into modular steps, each of which is methodologically acceptable both individually and in combination with the others.[16] It is characteristic of most simulations that this traditional transparency is unavailable. Is the presence of this kind of opacity necessarily a defect? To address this, consider that within models some loss of detail can enhance understanding. A well-known example is the century-old treatment of gases by statistical thermodynamics. The particular path consisting in the sequence of micro arrangements of molecules in the gas that the gas takes to thermal equilibrium has no intrinsic importance, and the system is best treated at a much higher level of description. Our ignorance of the omitted details is unimportant because the variables of maximal predictive value are the thermodynamical averages.[17] The same is true of many agent-based models in the social sciences and in biology. What is usually of interest is whether there are population-level equilibria, and the particular paths taken to arrive at those equilibrium points are of no great interest.

There are at least two sources of epistemic opacity. The first occurs when a computational process is too fast for humans to follow in detail. This is the situation with computationally assisted proofs of mathematical theorems such as the four-color theorem.[18] The second kind involves what Stephen Wolfram has called computationally irreducible processes,[19] processes that are of the kind best described, in David Marr's well-known classification, by Type 2 theories.[20] A Type 1 theory is one in which the algorithms for solving the theory can be treated as an issue separate from the computational problems that they are used to solve. A Type 2 theory is one in which a problem is solved by the simultaneous action of a considerable number of processes whose interaction is its own

---

[16]This point has been emphasized by Philip Kitcher (1989) as an alternative to Hempel's requirement that there merely exist a deductive relationship between the explanans and the explanandum. This is a very important issue to have highlighted, although for the record, I think it should be separated from the unification framework within which it appears.

[17]An interesting account of which level of description is predictively optimal for such systems can be found in Shalizi and Moore 2003.

[18]See, e.g., Tymoczko 1979; Teller 1980; Detlefsen and Luker 1980.

[19]Wolfram 1985; also Wolfram 2002, pp. 737–750.

[20]Marr 1982.

simplest description.[21] Epistemic opacity of this second kind plays a role in arguments that connectionist models of cognitive processes cannot provide an explanation of why or how those processes achieve their goals. The problem arises from the lack of an explicit algorithm linking the initial inputs with the final outputs, together with the inscrutability of the hidden units that are initially trained. Because computationally irreducible systems represented by Type 2 theories have been criticized by advocates of classical artificial intelligence[22] on the grounds that a proper understanding of how a system functions requires us to have an explicit algorithm at a level that correctly captures the structural aspects of the surface representation (i.e., at the 'conceptual' level, rather than what connectionists have called the 'subconceptual' level), we have here an analogue of the situation that we noted in section 4.6 holds for agent-based models, where no comprehensive model at the level of the total system exists. Computationally irreducible processes thus occur in systems in which the most efficient procedure for calculating the future states of the system is to let the system itself evolve.

Humphreys (1994) discusses in some detail how Ising models of ferromagnetism illustrate this second kind of kind of epistemic opacity, except that in the ferromagnetism case, there is an interesting combination of bottom-up Ising modeling and algorithmically driven Monte Carlo procedures. The reader is referred to that article for the details—the main point of relevance here is that the integrals involved in the Heisenberg version of the Ising model are unsolvable in any explicit way because they require us to integrate over an N-dimensional space where even in a much simplified model, N is of the order of 10^6 degrees of freedom. Although there exist analytic solutions to the Ising model for a one-dimensional lattice and for two-dimensional lattices with no external field, there is no such analytic treatment for other cases.[23] Analytically intractable integrals produce a standard context in which Monte Carlo methods of integration are used, and in this case they are supplemented by a sampling procedure known as the Metropolis algorithm[24] that is necessary to make the Monte Carlo method computationally feasible.

[21]See also Chaitin 1992.

[22]For example, Fodor and Pylyshyn 1988.

[23]See Tenius 1999. It is an interesting fact about such models that, provably, no phase transitions occur in the one-dimensional case with only short-range interactions.

[24]For details see Humphreys 1994.

To achieve its goal, the Metropolis algorithm constructs a random walk in configuration space having a limit distribution identical to that of the sampling function P. If this were merely a piece of abstract probability theory about Markov processes, it would, of course, be nothing exceptional. The key fact about this method is that the limit distribution is not computed by calculating a priori the values of the limit distribution. Rather, it is only by running concrete random walks, the details of which are never seen, that the limit distribution can be computed. A clear and transparent understanding of the details of the process between input and output is thus not available when a good fit is achieved between theory and data, and this is characteristic of many methods in the area.

We saw earlier that computational speed is an inescapable aspect of almost all methods of computational science, including those processes which are computationally reducible. When computational irreducibility is also present, the limits of human abilities become even more apparent. What counts as a computationally irreducible process is a subtle matter, but it is the existence of such processes that is the key issue for the problem of epistemic opacity. If we think in terms of such a process and imagine that its stepwise computation was slowed down to the point where, in principle, a human could examine each step in the process, the computationally irreducible process would become epistemically transparent. What this indicates is that the practical constraints we have previously stressed, primarily the need for computational speed, are the root cause of all epistemic opacity in this area. Because those constraints cannot be circumvented by humans, we must abandon the insistence on epistemic transparency for computational science. What replaces it would require an extended work in itself, but the prospects for success are not hopeless. The virtues of computational models—such as stability under perturbations of boundary conditions, scale invariance, conformity to analytic solutions where available, and so on—can be achieved by trial-and-error procedures treating the connections between the computational template and its solutions as a black box. There is nothing illegitimate about this as long as the motivation for the construction process is well founded. For example, one may have excellent reasons for holding that a particular parametric family of models is applicable to the case at hand, yet have only empirical methods available for deciding which parametric values are the right ones. In fact, the justification of scientific models on theoretical grounds rarely legitimates anything more than such a parametric family, and the adjustments in the fine structure by empirical data are simply an enhancement of the empirical input traditionally used to pin down the models more precisely.

5.4. Logical Form and Computational Form

Much of twentieth-century analytic philosophy has concerned itself with what is called the logical form of sentences (or propositions).[25] This focus on logical form has been fruitful and constitutes a powerful method for gaining insight into certain philosophical problems. Yet too intense a focus upon logical form can prove seriously misleading for certain purposes. Consider again the example of Newton's Second Law that was discussed in section 3.4. Two further consequences flow from this simple case.

The first is that if the equation representing Newton's Second Law is lawlike (and if it is not, what is?), then it would be an error to think of equations (3.2) and (3.3) as substitution instances of (3.1), for what constitutes the crucial difference in form between (3.2) and (3.3) is the linear versus nonlinear property of the "substitution instance" of the general schema (3.1), and not the preservation of its second-order form. Indeed, there are "substitution instances" of (3.1) that would not even preserve that second-order property if, for example, one had a force dependent upon a third-order derivative. Similarly, "substitution instances" of (3.1) could be either homogeneous or inhomogeneous, a difference that would also have an important bearing on their solvability.[26] Clearly, these issues can be treated by careful attention to invariance of form, but it is not the logical form that is important here.

In the case we have considered, this means that we should not represent the template at a level of abstraction which would destroy the mathematical form of the equation. To be clear about this: One could reformulate equations (3.1), (3.2), and (3.3) in a standard logical language

---

[25]This has been true at least since Bertrand Russell's pathbreaking 1905 article "On Denoting." One of Russsell's important contributions in that article was to show that the true underlying logical form of a sentence can be quite different from its superficial grammatical appearance.

[26]An analogy here might be the claim that $\forall x\, \exists y\, Fxyz$, $\forall x\, \forall y\, Fxyz$ are both substitution instances of the logical scheme $\forall \alpha \phi$, thus sliding over the fact that when F is a recursive relation, these occupy different places in the arithmetic hierarchy, the first being π_2^0 and the second (via contraction of quantifiers), π_1^0. Since by Kleene's Arithmetical Hierarchy Theorem, there are π_n^0 unary predicates that are not π_m^0 for $m < n$, the simple scheme is a misrepresentation of logical form.

by using variable-binding operators,[27] thus forcing them into the universally quantified conditional form beloved of philosophical textbooks. But why do it? As we argued throughout chapters 3 and 4, what is important is not solvability or computability in principle, which has usually been the motivation behind metamathematical investigations, but the actual ability to get general or specific solutions to the equations. This depends directly upon the explicit mathematical form rather than upon the underlying logical form, and means dealing with the theories at a lower level of abstraction than is appropriate in investigating abstract provability relations. In fact, we can generalize this and say that it is the general computational form of the theory that counts, where the syntactic form of the theory now includes its formulation in some specific computational language. Standard mathematical languages are oriented toward human computational devices, but we should not take that as the sole arbiter of 'computational form'.

Let us consider just one defect of using logic as the primary tool in philosophy of science. Logic, whether construed formally or platonistically, is atemporal. Construed intuitionistically, it has a distinctive temporal element but also an overtly psychological aspect. So to treat dynamical phenomena in science, logical reconstructions must reconstruct the temporal dimension in an atemporal format. This is standardly done within a B-theory conception of time which denies that there is any need to impute a genuinely dynamic aspect to time.[28] Yet, as we have seen, two of the key features of computer simulations are their essentially dynamic aspect—turbulent flow can be shown as it develops and persists, for example, or the evolution of galaxies can be shown on a vastly compressed timescale—and the presence of the internal clock in the hardware. A re-representation of those features in an atemporal form would simply omit what is distinctive about simulations.

Matter constrains method. By this I mean that the intrinsic nature of a given phenomenon renders certain methods and forms of inquiry

---

[27]Differential equations can be formalized within standard logical languages by using this technique. See, for example, chapter 11 in Kalish et al. 1980; also Montague et al. unpublished. However, the point remains that this kind of formalization takes one away from what is important about the form of the equations, and hence is at best unnecessary; at worst it distorts the important formal features.

[28]The B-theory of time holds that the relation of "earlier than," which imposes a total ordering on points of time, is sufficient to account for all aspects of time. For one survey of this view, see Le Poidevin 1998.

impotent for that subject, whereas other subject matters will yield to those means. Heating plays a key role in polymerase chain reaction processes, but virtually none in studying radioactive decay; randomized controlled trials are the gold standard in epidemiological investigations, but are unnecessary for electromagnetic experiments; careful attention to statistical analysis is required in mouse models of carcinogenesis, but such analysis is often casual window dressing in physics.[29] Subject matter does not determine method—too many other factors enter for that to be true—but the philosophical dream for much of the twentieth century was that logical, or at least formal, methods could, in a subject-matter-independent way, illuminate some, if not all, of scientific method used to reach the five goals above. Although these methods achieved impressive results and research continues in that tradition, there are quite serious limitations on the extent to which those methods are capable of shedding light on our goals.

Like most all-purpose tools, logic is not well suited to deal with every goal of science, or even particularly well with any one of them. In saying this, I am not in any way denigrating the remarkable achievements of twentieth-century logic, nor the clarificatory role it has played in the philosophy of science. Most attacks on logically based methods are based on a quite remarkable combination of hostility and ignorance,[30] and I want specifically to disassociate myself from those.

5.5 In Practice, Not in Principle

Much of philosophy, and of philosophy of science, is concerned with what can be done in principle. Philosophy is as entitled as any enterprise to use idealizations where they are appropriate, and "in principle" arguments are legitimate when one is concerned with impossibility results; what cannot be done in principle, can't be done in practice either. And so negative results and impossibility theorems have played a central role in twentieth-century philosophy: Gödel's Incompleteness Theorems, Heisenberg's uncertainty relations, Bell's Theorem, Arrow's Impossibility Theorem, the impossibility of material particles exceeding the speed of light, Pauli's Exclusion Principle, and many others.

---

[29]For evidence on the last, see Humphreys 1976, appendix.

[30]For a couple of typical examples, see Nye 1990; Collins 1984.

Yet when one is concerned with positive rather than negative results, the situation is reversed—what can be done in principle is frequently impossible to carry out in practice. Of course philosophy should not ignore these 'in principle' results, but if we are concerned with, among other things, how science progresses, then the issue of how science pushes back the boundaries of what can be known in practice should be a primary concern. That is because scientific progress involves a temporally ordered sequence of stages, and one of the things which influences that progress is that what is possible in practice at one stage was not possible in practice at an earlier stage.[31] If one focuses on what is possible in principle (i.e., possible in principle according to some absolute standard rather than relative to constraints that are themselves temporally dependent), this difference cannot be represented, because the possibility-in-principle exists at both stages of development (and indeed exists eternally). To say that for scientific purposes a function is computable when it is computable only in principle and not in practice, is rather like presenting a friend with a million dollars inside a safe sealed with an uncrackable lock and telling him, "Now you are a millionaire."

There is one 'in principle' situation that tends to play a large role in the philosophy of science, especially in discussions of realism, and that is an appeal to what is called 'limit science'. This is taken to be the state of science at the end of scientific inquiry, when a final set of theories is available that satisfies all the desiderata of science, including compatibility with all the evidence. It is then often claimed that there will always be alternative theories consistent with the evidence, "even when all the evidence is in."[32] There is one clear sense intended here—when we have all the evidence that can be represented in some standard language; but this is far too limiting. We need to know more about what 'all the evidence' means. All the evidence according to our current standards of evidence? Surely not, for what counts as evidence has changed over the centuries. All possible evidence? Surely not, for then we would be in the position of the epistemic god described in the first section of chapter 1, and that god had no use for science. And possible according to which

---

[31]I had earlier, in a very limited sense, suggested that we observe this difference (Humphreys 1995, pp. 500–501), but the importance of approaching philosophical issues from an "in practice" perspective was impressed upon me after many discussions with Bill Wimsatt. Some of his reasons for adopting this orientation can be found in Wimsatt 2004.

[32]See, for example, Putnam 1980.

modality? Not all logically possible evidence. Perhaps all nomologically possible evidence, but more likely, all scientifically possible evidence. And if what counts as evidence depends upon what the current state of science enables us to detect using technology based on that science, the very concept of the evidence available to limit science becomes partially circular unless we have some argument that all possible historical developments of science would end with the same limit science. Absent that argument, what limit science would be like, even if we assume that such a thing exists, is so indeterminate that without further specification it is not very useful as the basis of philosophical arguments. In the end, such questions do not affect the underdetermination results, because they are strong enough to hold under all interpretations of 'all possible', but this very abstractness indicates that the connection with real science is extremely tenuous. We could be dealing with limit musicology and encounter the same underdetermination. The arguments are not, despite their mode of presentation, about *scientific* realism.

In order to avoid misunderstanding, I should make it clear that shifting from what is possible in principle to what is possible in practice does not preclude us from maintaining a distinction between the normative and the descriptive. Much attention has been focused on the fact that because humans have a limited cognitive apparatus, normative ideals of rationality are not always satisfied and, in consequence, humans employ heuristic strategies that occasionally deviate, sometimes radically, from the accepted norms of rational evaluation.[33] It is a mistake to draw the conclusion from these descriptive studies that normative considerations have no place in human reasoning and that descriptive studies should be given priority. The goal should not be to limit oneself to what unaided human reasoning ability can achieve, but to discover what standards of rationality can be achieved when our native talents are supplemented by the theoretical and computational devices available to us through research. For example, the parallels that are often drawn between the need to resort to heuristics in finding a suboptimal solution to the traveling salesman problem and satisficing methods in human problem solving (i.e., finding suboptimal but adequate solutions)[34] are mistaken. Just as it would be foolish to limit what was combinatorially possible in practice for a given traveling salesman problem to what could be counted

---

[33] See, e.g., Kahnemann et al. 1982; Cohen 1981; Gigerenzer et al. 1999.
[34] See Holyoak 1990.

with the aid of pencil and paper, so it is unnecessarily constraining to preclude the use of theories of rationality simply because unaided human computational abilities find them hard to implement.[35] One of the advantages of science is that we can recognize our own limitations and faced, for example, with a problem of statistical inference, we can avail ourselves of highly sophisticated and reliable techniques rather than relying on our own, highly fallible, approaches. Scientific reasoning is not everyday reasoning.

5.6 Conclusion

When Protagoras suggested in the fifth century B.C. that man is the measure of all things, the narrowness of his epigram could be excused. Now, perhaps, it is only solipsists who can claim for themselves that peculiarly privileged position. The Copernican Revolution first removed humans from their position at the center of the physical universe, and science has now driven humans from the center of the epistemological universe. This is not to be lamented, but to be celebrated. There is one final, minor, conclusion. The philosophy of science, or at least that part of it which deals with epistemology, no longer belongs in the humanities.

---

[35]Some writers in the philosophy of mathematics have explored these differences: "An undecided statement, S, is decidable in principle just in case an appropriately large but finite extension of our capacities would confer on us the ability to verify or falsify it in practice" (Wright 1987, p. 113). Quoted by Charles Parsons in "What Can We Do 'In Principle,'" talk at Tenth International Congress of LMPS, Florence, August 20, 1995.

References

Abbey, D., Hwang, B., Burchette, R., Vancuren, T., and Mills, P. 1995. "Estimated Long-Term Ambient Concentrations of PM_{10} and Development of Respiratory Problems in a Nonsmoking Population." *Archives of Environmental Health* 50, 139–152.

Achinstein, P. 1968. *Concepts of Science*. Baltimore: Johns Hopkins University Press.

Adams, R.M. 1979. "Primitive Thisness and Primitive Identity." *Journal of Philosophy* 76, 5–26.

Adelman, S., Dukes, R., and Adelman, C. 1992. *Automated Telescopes for Photometry and Imaging*. Chelsea, Mich.: Astronomical Society of the Pacific.

Alston, W. 1998. "Empiricism." In *The Routledge Encyclopedia of Philosophy*, vol. 3. London: Routledge.

Armstrong, D. 1978. *A Theory of Universals*. Cambridge: Cambridge University Press.

———. 1989. *Universals: An Opinionated Introduction*. Boulder, Colo.: Westview Press.

———. 1997. *A World of States of Affairs*. Cambridge: Cambridge University Press.

Arthur, W.B., Holland, J., LeBaron, B., Palmer, R., and Taylor, P. 1997. "Asset Pricing Under Endogenous Expectations in an Artificial Stock Market." In W.B. Arthur, S. Durland, and D. Lane, eds., *The Economy as an Evolving Complex System, vol. 2*, 15–44. Reading, Mass.: Addison-Wesley.

Axelrod, R. 1997. *The Complexity of Cooperation*. Princeton, N.J.: Princeton University Press.

Azzouni, J. 1997. "Thick Epistemic Access: Distinguishing the Mathematical from the Empirical." *Journal of Philosophy* 94, 472–484.

Balzer, W., Moulines, C.U., and Sneed, J. 1987. *An Architectonic for Science*. Dordrecht: D. Reidel.

Barr, D., and Zehna, P. 1971. *Probability*. Belmont, Calif.: Brooks/Cole.

Barwise, J., and Etchemendy, J. 1992. *The Language of First Order Logic*. Stanford, Calif.: CSLI.

Barwise, J., and Feferman, S. 1985. *Model Theoretic Logics*. New York: Springer-Verlag.

Binder, K., and Heerman, D.W. 1988. *Monte Carlo Simulation in Statistical Physics*. Berlin: Springer-Verlag.

Bogen, J., and Woodward, J. 1988. "Saving the Phenomena." *Philosophical Review* 97, 303–352.

Bridgman, P. 1927. *The Logic of Modern Physics*. New York: Macmillan.

Brummell, N., Cattaneo, F., and Toomre, J. 1995. "Turbulent Dynamics in the Solar Convection Zone." *Science* 269, 1370.

Campbell, K. 1990. *Abstract Particulars*. Oxford: Basil Blackwell.

Cargile, J. 1972. "In Reply to 'A Defense of Skepticism.'" *Philosophical Review* 81, 229–236.

———. 1987. "Definitions and Counterexamples." *Philosophy* 62, 179–193.

Cartwright, N. 1983. *How the Laws of Physics Lie*. Oxford: Oxford University Press.

Chaitin, G. 1992. *Algorithmic Information Theory*. Cambridge: Cambridge University Press.

Chang, H. 2004. *The Philosophical Thermometer: Measurement, Metaphysics, and Scientific Progress*. Oxford: Oxford University Press.

Christmann, J. 1612. *Nodus Gordius ex doctrina sinnum explicatus*. Heidelberg.

Churchland, P. 1985. "The Ontological Status of Observables: In Praise of the Super-empirical Virtues." In P. Churchland and C. Hooker, eds., *Images of Science: Essays on Realism and Empiricism*, Chap. 2. Chicago: University of Chicago Press.

Cipra, B. 1995. "Mathematicians Open the Black Box of Turbulence." *Science* 269, 1361–1362.

Clark, A. 1989. *Microcognition*. Cambridge, Mass.: MIT Press.

Cohen, L.J. 1981. "Can Human Irrationality Be Experimentally Demonstrated?" *Behavioral and Brain Sciences* 4, 317–331.

Cohen, N.C. 1996. "The Molecular Modeling Perspective in Drug Design." In N.C. Cohen, ed., *Guidebook on Molecular Modeling in Drug Design*, 1–17. San Diego: Academic Press.

Collins, H.M. 1985. *Changing Order*. London: Sage.

Collins, R. 1984. "Statistics Versus Words." In R. Collins, ed., *Sociological Theory 1984*, 329–362. San Francisco: Jossey-Bass.

Comte, A. 1869. *Cours de philosophie positive*, 3rd ed., vol. 1. Paris: J.B. Bailliere et Fils.

Conte, R., Hegselman, R., and Terna, P., eds. 1997. *Simulating Social Phenomena*. New York: Springer-Verlag.

Crosson, F., and Sayre, K. 1963. "Simulation and Replication." In K. Sayre and F. Crosson, eds., *The Modeling of Mind*, 3–24. Notre Dame, Ind.: University of Notre Dame Press.

Darius, J. 1984. *Beyond Vision*. Oxford: Oxford University Press.

Denef, J., and Lipshitz, L. 1984. "Power Series Solutions of Algebraic Differential Equations." *Mathematische Annalen* 267, 213–238.

Denef, J., and Lipshitz, L. 1989. "Decision Problems for Differential Equations." *Journal of Symbolic Logic* 54, 941–950.

DeRose, K. 1995. "Solving the Skeptical Problem." *Philosophical Review* 104, 1–52.

Detlefsen, M., and Luker, M. 1980. "The Four Color Theorem and Mathematical Proof." *Journal of Philosophy* 77, 803–819.

Drake, S. 1970. "Galileo and the Telescope." In S. Drake, *Galileo Studies*, 140–158. Ann Arbor: University of Michigan Press.

Duhem, P. 1906. *The Aim and Structure of Physical Theory*. English translation by Philip Wiener. Princeton, N.J.: Princeton University Press, 1991.

Earman, J. 1986. *A Primer on Determinism*. Dordrecht: D. Reidel.

Elder, C. 2001. "Contrariety and the Individuation of Properties." *American Philosophical Quarterly* 38, 249–260.

Epstein, J. 1997. *Non-linear Dynamics, Mathematical Biology, and Social Science*. Reading, Mass.: Addison-Wesley.

Epstein, J., and Axtell, R. 1996. *Growing Artificial Societies: Social Science from the Bottom Up*. Cambridge, Mass.: MIT Press.

Feller, W. 1968a. *An Introduction to Probability Theory and Its Applications*, vol. 1, 3rd ed. New York: John Wiley.

———. 1968b. *An Introduction to Probability Theory and Its Applications*, vol. 2, 2nd ed. New York: John Wiley.

Feynman, R. 1965. *The Feynman Lectures on Physics*, by R. Feynman, R. Leighton, and M. Sands, vol. 2. Reading, Mass: Addison-Wesley.

———. 1985. *Surely You're Joking, Mr. Feynman*. New York: W.W. Norton.

Fodor, J., and Pylyshyn, Z. 1988. "Connectionism and Cognitive Architecture: A Critical Analysis." *Cognition* 28, 3–71.

Franklin, A. 1986. *The Neglect of Experiment*. Cambridge: Cambridge University Press.

———. 1990. *Experiment, Right or Wrong*. Cambridge: Cambridge University Press.

———. 1997. "Calibration." *Perspectives on Science* 5, 31ff.

Freedman, D. 1985. "Statistics and the Scientific Method." In W. Mason and S. Fienberg, eds., *Cohort Analysis in Social Research: Beyond the Identification Problem*, 343–366. New York: Springer-Verlag.

Freedman, D., and Zeisel, H. 1988. "From Mouse to Man: The Quantitative Assessment of Cancer Risks." *Statistical Science* 3, 3–56 (with responses and replies).

Freidhoff, R.M., and Benzon, W. 1989. *Visualization: The Second Computer Revolution*. New York: Harry Abrams.

French, A.P., and Taylor, E.F. 1998. *An Introduction to Quantum Physics*. Cheltenham, U.K.: Stanley Thomas.

Friedman, M. and Demopoulos, W. 1989. "The Concept of Structure in 'The Analysis of Matter.'" In C. Wade Savage, ed., *Rereading Russell*, 183–199. Minneapolis: University of Minnesota Press.

Galilei, Galileo. 1623. *The Assayer*. In *Discoveries and Opinions of Galileo*. Translated with introduction and notes, by Stillman Drake. Garden City, N.Y.: Doubleday, 1957.

Gamble, J. 1998. "$PM_{2.5}$ and Mortality in Long-term Prospective Cohort Studies: Cause–Effect or Statistical Associations?" *Environmental Health Perspectives* 106, 535–549.

Gamble, J., and Lewis, R. 1996. "Health and Respirable Particulate (PM) Air Pollution: A Causal or Statistical Association?" *Environmental Health Perspectives* 104, 838–850.

Gigerenzer, G., Todd, P., and the ABC Research Group. 1999. *Simple Heuristics That Make Us Smart*. New York: Oxford University Press.

Goldstein, H. 1959. *Classical Mechanics*. Cambridge, Mass.: Addison-Wesley.

Goodman J., Narayan, R., and Goldreich, P. 1987. "The Stability of Accretion Tori—II. Non-linear Evolution to Discrete Planets." *Monthly Notices of the Royal Astronomical Society* 225, 695.

Goodman, Nelson. 1977. *The Structure of Appearance*, 3rd ed. Dordrecht: D. Reidel.

Gould, H., and Tobochnik, J. 1988. *An Introduction to Computer Simulation Methods, Parts 1 and 2*. Reading, Mass.: Addison-Wesley.

Hacking, Ian. 1981. "Do We See Through a Microscope?" *Pacific Philosophical Quarterly* 62, 305–322.

———. 1983. *Representing and Intervening*. Cambridge: Cambridge University Press.

Hall, A.R. 1983. *The Revolution in Science 1500–1750*. New York: Longman.

Hall, J.A. 1966. *The Measurement of Temperature*. London: Chapman and Hall.

Hammer, R., Hocks, M., Kulish, U., and Ratz, D. 1995. *C++ Toolbox for Verified Computing*. Berlin: Springer-Verlag.

Hartmann, S. 1996. "The World as a Process: Simulation in the Natural and Social Sciences." In R. Hegselmann, U. Muller, and K. Troitzsch, eds., *Modelling and Simulation in the Social Sciences from the Philosophy of Science Point of View*, 77–100. Dordrecht: Kluwer Academic.

Harvey, D., Gregory, J., et al. 1997. "An Introduction to Simple Climate Models Used in the IPCC Second Assessment Report." Intergovernmental Panel on

Climate Control Technical Paper II. Available at http://www.wmo.ch/indexflash.html.

Hasslacher, B. 1993. "Parallel Billiards and Monster Systems." In N. Metropolis and G.-C. Rota, eds., *A New Era in Computation*, 53–66. Cambridge, Mass.: MIT Press.

Hawley, J. 1987. "Non-linear Evolution of a Non-axisymmetric Disc Instability." *Monthly Notices of the Royal Astronomical Society* 225, 667.

———. 1988. "Numerical Simulation of AGNS." *Advances in Space Research* 8, 119–126.

———. 1995. "Keplerian Complexity: Numerical Simulations of Accretion Disk Transport." *Science* 269, 1365.

Hedström, P., and Swedberg, R., eds. 1998. *Social Mechanisms: An Analytical Approach to Social Theory*. Cambridge: Cambridge University Press.

Hegselmann, R., Mueller, U., and Troitzsch, K., eds. 1996. *Modelling and Simulation in the Social Sciences from the Philosophy of Science Point of View*. Dordrecht: Kluwer Academic.

Hempel, C. 1954. "A Logical Appraisal of Operationalism." *Scientific Monthly* 79, 215–220. Reprinted in C. Hempel, *Aspects of Scientific Explanation and Other Essays*. New York: Free Press, 1965.

Henry, G.W., and Eaton, J.A. 1995. *Robotic Telescopes: Current Capabilities, Present Developments, and Future Prospects for Automated Astronomy. Proceedings of a Symposium Held as Part of the 106th Annual Meeting of the Astronomical Society of the Pacific, Flagstaff, Arizona, 28–30 June 1994*. San Francisco: Astronomical Society of the Pacific.

Heppenheimer, T.A. 1991. "Some Tractable Mathematics for Some Intractable Physics." *Mosaic* 22, 29–39.

Hesse, M. 1966. *Models and Analogies in Science*. Notre Dame, Ind.: University of Notre Dame Press.

Hilbert, D. 1926. "On the Infinite." Reprinted in P. Benacerraf and H. Putnam, eds., *Philosophy of Mathematics: Selected Readings*, 2nd ed., 183–201. Cambridge: Cambridge University Press 1983. English translation of original in *Mathematische Annalen* 95 (1926), 161–190.

Hill, B.C., and Hinshaw, W.S. 1985. "Fundamentals of NMR Imaging." In *Three Dimensional Biomedical Imaging*, vol. 2, R.A. Robb, ed., 79–124. Boca Raton, Fla.: CRC Press.

Hodges, A. 1983. *Alan Turing: The Enigma*. New York: Simon and Schuster.

Holland, P. 1993. "Complex Adaptive Systems." In N. Metropolis and G.-C. Rota, eds., *A New Era in Computation*, 17–30. Cambridge, Mass.: MIT Press.

Holyoak, K. 1990. "Problem Solving." In D. Osherson and E. Smith, eds., *Thinking: An Invitation to Cognitive Science*, vol. 3, 117–146. Cambridge, Mass.: MIT Press.

Hopfield, J., and Tank, D. 1986. "Computing with Neural Circuits: A Model." *Science* 233, 625–633.

Hubbard, Roderick E. 1996. "Molecular Graphics and Modelling: Tools of the Trade." In N.C. Cohen, ed., *Guidebook on Molecular Modeling in Drug Design*, 19–54. San Diego: Academic Press.

Humphreys, P. 1976. "Inquiries in the Philosophy of Probability: Randomness and Independence." Ph.D. dissertation, Stanford University.

———. 1989. *The Chances of Explanation: Causal Explanation in the Social, Medical and Physical Sciences*. Princeton, N.J.: Princeton University Press.

———. 1991. "Computer Simulations." In A. Fine, M. Forbes, and L. Wessels, eds., *PSA 1990*, vol. 2, 497–506. East Lansing, Mich.: Philosophy of Science Association.

———. 1993a. "Greater Unification Equals Greater Understanding?" *Analysis* 53, 183–188.

———. 1993b. "Seven Theses on Thought Experiments." In J. Earman, A. Janis, G. Massey, and N. Rescher, eds., *Philosophical Problems of the Internal and External World*, 205–228. Pittsburgh, Pa.: University of Pittsburgh Press.

———. 1994. "Numerical Experimentation." In P. Humphreys, ed., *Patrick Suppes, Scientific Philosopher*, vol. 2, 103–118. Dordrecht: Kluwer Academic.

———. 1995. "Computational Science and Scientific Method." *Minds and Machines* 5, 499–512.

———. 2000. "Analytic Versus Synthetic Understanding." In J. H. Fetzer, ed., *Science, Explanation, and Rationality: The Philosophy of Carl G. Hempel*, 267–286. Oxford: Oxford University Press.

Huntsberger, D., and Billingsley, P. 1973. *Elements of Statistical Inference*, 3rd ed. Boston: Allyn and Bacon.

Iverson, G., et al. 1971. "Biases and Runs in Dice Throwing and Recording: A Few Million Throws." *Psychometrika* 36, 1–19.

Jackson, J.D. 1962. *Classical Electrodynamics*. New York: John Wiley.

Jaśkowski, S. 1954. "Example of a Class of Systems of Ordinary Differential Equations Having No Decision Method for Existence Problems." *Bulletin de l'Académie Polonaise des Sciences, classe* 3, 2, 155–157.

Jensen, H. J. 1998. *Self-Organized Criticality: Emergent Complex Behavior in Physical and Biological Systems*. Cambridge: Cambridge University Press.

Kahneman, D., Slovic, P., and Tversky, A. 1982. *Judgement Under Uncertainty: Heuristics and Biases*. Cambridge: Cambridge University Press.

Kalish, D., Montague, R., and Mar, G. 1980. *Logic: Techniques of Formal Reasoning*. New York: Harcourt, Brace.

Karplus, W. 1958. *Analog Simulation: Solution of Field Problems*. New York: McGraw-Hill.

Kaufmann, W.J., and Smarr, L. 1993. *Supercomputing and the Transformation of Science*. New York: W.H. Freeman.

Kibble, T.W.B. 1966. *Classical Mechanics*. New York: McGraw-Hill.

Kitcher, P. 1984. *The Nature of Mathematical Knowledge*. Oxford: Oxford University Press.

———. 1989. "Explanatory Unification and the Causal Structure of the World." In P. Kitcher and W. Salmon, eds., *Scientific Explanation*. Minnesota Studies in the Philosophy of Science, 13. Minneapolis: University of Minnesota Press.

———. 2001. *Science, Truth, and Democracy*. Oxford: Oxford University Press.

Kline, M. 1972. *Mathematical Thought from Ancient to Modern Times*. New York: Oxford University Press.

Kohler, T., and Gumerman, G. 1999. *Dynamics in Human and Primate Societies*. Oxford: Oxford University Press.

Kolmogorov, A.N. 1942. "Equations of Turbulent Motion in an Incompressible Fluid." *Izv. Akad. Nauk S.S.S.R.*, Ser. Fiz. 6, 56. English translation, Imperial College, London, Mechanical Engineering Department, Report ON/6, 1968.

Kuhn, T. 1970. *The Structure of Scientific Revolutions*, 2nd ed. Chicago: University of Chicago Press.

———. 1977. "A Function for Thought Experiments." In T. Kuhn, *The Essential Tension*, 240–265. Chicago: University of Chicago Press.

Lakatos, I. 1970. "Falsification and the Methodology of Scientific Research Programs." In I. Lakatos and A. Musgrave, *Criticism and the Growth of Knowledge*, 91–196. Cambridge: Cambridge University Press.

Laplace, P.S. 1825. *A Philosophical Essay on Probabilities*. English translation of 5th French ed. New York: Springer-Verlag, 1995.

Laymon, R. 1982. "Scientific Realism and the Hierarchical Counterfactual Path from Data to Theory." In P.D. Asquith and Thomas Nickles, eds., *PSA 1982: Proceeding of the 1982 Biennial Meeting of the Philosophy of Science Association*, vol. 1, 107–121. East Lansing, Mich.: Philosophy of Science Association.

Le Poidevin, R. 1998. "The Past, Present, and Future of the Debate About Tense." In R. Le Poidevin, ed., *Questions of Time and Tense*, 13–42. Oxford: Oxford University Press.

Lehrer, K. 1995. "Knowledge and the Trustworthiness of Instruments." *The Monist* 78, 156–170.

Levin, Simon. 1980. "Mathematics, Ecology, and Ornithology" *Auk* 97, 422–425.

Lewis, D. 1986. *On the Plurality of Worlds*. Oxford: Basil Blackwell.

Marr, D. 1982. *Vision*. Cambridge, Mass.: MIT Press.

Maxwell, G. 1962. "The Ontological Status of Theoretical Entities." In H. Feigl and G. Maxwell, eds., *Minnesota Studies in the Philosophy of Science*, vol. 3, 3–15. Minneapolis: University of Minnesota Press.

McWeeny, R., and Sutcliffe, B. 1969. *Methods of Molecular Quantum Mechanics*. New York: Academic Press.

Mendelson, E. 1987. *Introduction to Mathematical Logic*, 3rd ed. Belmont, Calif.: Wadsworth.

Metropolis, N. 1993: "The Age of Computing: A Personal Memoir." In N. Metropolis and G.-C. Rota, eds., *A New Era in Computation*, 119–130. Cambridge, Mass.: MIT Press.

Mezard, M., Parisi, G., and Virasoro, M. 1987. *Spin Glass Theory and Beyond*. Singapore: World Scientific.

Mihram, G. 1972. *Simulation: Statistical Foundations and Methodology*. New York: Academic Press.

Mill, J.S. 1874. *A System of Logic*, 8th ed., London: Longman, Green.

Montague, R. 1962. "Deterministic Theories." *Decision, Theories, and Groups* 2, 325–370. Reprinted in R. Montague, *Formal Philosophy*, 303–359. New York: Yale University Press, 1974.

Montague, R., Scott, D., and Tarski, A. "An Axiomatic Approach to Set Theory." unpublished.

Moor, W.H., Grant, D.G., and Pavis, M.R. 1993. "Recent Advances in the Generation of Molecular Diversity." *Annual Reports in Medical Chemistry* 28, 315–324.

Morgan, M., and Morrison, M. 1999. *Models as Mediators*. Cambridge: Cambridge University Press.

Morse, P., and Feshbach, H. 1953. *Methods of Theoretical Physics*, part 1. New York: McGraw-Hill.

Newman, M. 1928. "Mr Russell's Causal Theory of Perception." *Mind* 37, 143ff.

Nieuwpoort, W.C. 1985. "Science, Simulation, and Supercomputers." In J. Devreese and P. van Camp, eds., *Supercomputers in Theoretical and Experimental Science*, 3–9. New York: Plenum Press.

Nye, A. 1990. *Words of Power*. New York: Routledge.

Ord-Smith, R.J., and Stephenson, J. 1975. *Computer Simulation of Continuous Systems*. Cambridge: Cambridge University Press.

Ortega, J., and Poole, W., Jr. 1981. *An Introduction to Numerical Methods for Differential Equations*. Marshfield, Mass.: Pitman.

Oxbury, H. 1985. *Great Britons: Twentieth Century Lives*. Oxford: Oxford University Press.

Pearl, J. 2000. *Causality*. Cambridge: Cambridge University Press.

Petersen, A. 2000. "Philosophy of Climate Science." *Bulletin of the American Meteorological Society* 81, 265–271.

Poincaré, H. 1902. *La Science et hypothèse*. Paris: E. Flammarion.

———. 1952. *Science and Hypothesis*. New York: Dover Books.

Press, W.H., et al. 1988. *Numerical Recipes in C: The Art of Scientific Computing*. New York: Cambridge University Press.

Putnam, H. 1980. "Models and Reality." *Journal of Symbolic Logic* 45, 464–482.

Quine, W.V.O. 1951. "Two Dogmas of Empiricism." *Philosophical Review* 60, 20–43. Reprinted in Quine, *From a Logical Point of View*, 20–46.

———. 1961a: "Logic and the Reification of Universals." In Quine, *From a Logical Point of View*, 102–129.

———. 1961b. "Identity, Ostension, and Hypostasis." In Quine, *From a Logical Point of View*, 65–79.

———. 1961c. From a Logical Point of View, 2nd rev., ed. Cambridge, Mass.: Harvard University Press.

Redhead, M. 1980. "Models in Physics." *British Journal for the Philosophy of Science* 31, 145–163.

Richtmyer, R.D. and Morton, K.W. 1967. *Difference Methods for Initial-Value Problems*. 2nd ed. New York: Interscience.

Robinson, R. 1968. *Definition*. Oxford: Clarendon Press.

Rohrlich, F. 1991. "Computer Simulations in the Physical Sciences." In A. Fine, M. Forbes, and L. Wessels, eds., *PSA 1990*, vol. 2, 507–518. East Lansing, Mich.: Philosophy of Science Association.

Rosenbluth, A., and Weiner, N. 1945. "The Role of Models in Science." *Philosophy of Science* 12, 316–321.

Rubel, L. 1989. "Digital Simulation of Analog Computation and Church's Thesis." *Journal of Symbolic Logic* 54, 1011–1017.

Russ, John C. 1990. *Computer Assisted Microscopy*. New York: Plenum Press.

Russell, B. 1905. "On Denoting." *Mind* 14, 479–493.

Sayre, K. 1965. *Recognition*. Notre Dame, Ind.: University of Notre Dame Press.

Sayre, K., and Crosson, F., eds. 1963: *The Modeling of Mind*. Notre Dame, Ind.: University of Notre Dame Press.

Schelling, T. 1978. *Micromotives and Macrobehavior*. New York: Norton.

Schooley, J.F. 1982. *Temperature: Its Measurement and Control in Science and Industry*, vol. 5, part 2. New York: American Institute of Physics.

Shalizi, C. and Moore, C. 2003. "What Is a Macrostate? Subjective Observations and Objective Dynamics." http://philsci-archive.pitt.edu, document no. PITT-PHIL-SCI00001119.

Shin, S.-J. 1994: *The Logical Status of Diagrams*. Cambridge: Cambridge University Press.

Skyrms, B. 1996. *The Evolution of the Social Contract*. Cambridge: Cambridge University Press.

Smarr, L. 1985. "An Approach to Complexity: Numerical Computations." *Science* 228, 403–408.

Sochurek, Howard. 1988. *Medicine's New Vision*. Eastam, Pa.: Mack.

Sterken, C., and Manfroid, J. 1992. *Astronomical Photometry: A Guide*. Dordrecht: Kluwer Academic.

Stroud, B. 1984. *The Significance of Philosophical Scepticism*. Oxford: Clarendon Press.

Suppes, P. 1960. "A Comparison of the Meaning and Uses of Models in Mathematics and the Empirical Sciences." *Synthèse* 12, 287–300.

———. 1962. "Models of Data." In E. Nagel et al., eds., *Logic, Methodology, and Philosophy of Science: Proceedings of the 1960 International Congress*, 252–261. Stanford, Calif.: Stanford University Press.

———. 1970. "Set-Theoretic Structures in Science." Stanford, Calif.: Institute for Mathematical Studies in the Social Sciences (mimeoed manuscript). Published with additions as *Representation and Invariance of Scientific Structures*. Stanford, Calif.: CSLI, 2002.

Tajima, T. 1989. *Computational Plasma Physics*. Redwood City, Calif.: Addison-Wesley.

Tannehill, J.C., Anderson, D.A., and Pletcher, R.H. 1997. *Computational Fluid Mechanics and Heat Transfer*, 2nd ed. Washington, D.C.: Taylor and Francis.

Tarski, A. 1956. "The Concept of Truth in Formalized Languages." In A. Tarski, *Logic, Semantics, Metamathematics*, 153–278. Oxford: Clarendon Press.

Taylor, J.G. 1975. *Superminds: An Inquiry into the Paranormal*. London: Macmillan.

Teller, P. 1980. "Computer Proof." *Journal of Philosophy* 77, 797–803.

Tenius, D. 1999. *Statistical Mechanics*. Bristol, U.K.: Institute of Physics.

Turing, A. 1950. "Computing Machinery and Intelligence." *Mind* 59, 433–460.

Tymoczko, T. 1979. "The Four-Color Problem and Its Philosophical Significance." *Journal of Philosophy* 76, 57–83.

Van Fraassen, B. 1980. *The Scientific Image*. Oxford: Clarendon Press.

———. 1985. "Empiricism in the Philosophy of Science." In P. Churchland and C. Hooker, eds., *Images of Science: Essays on Realism and Empiricism*, 245–308. Chicago: University of Chicago Press.

———. 1989. *Laws and Symmetry*. Oxford: Clarendon Press.

———. 2002. *The Empirical Stance*. New Haven, Conn.: Yale University Press.

Van Helden, A. 1974. "The Telescope in the Seventeenth Century." *Isis* 65, 38–58. Reprinted in P. Dear, ed., *The Scientific Enterprise in Early Modern Europe*. Chicago: University of Chicago Press, 1997.

———. 1977. *The Invention of the Telescope*. Transactions of the American Philosophical Society 67, part 4. Philadelphia: American Philosophical Society.

Westfall, R. 1980. *Never at Rest: A Biography of Isaac Newton*. Cambridge: Cambridge University Press.

Whewell, W. 1840. *The Philosophy of the Inductive Sciences*, vol. 2. London: J.W. Parker.

Wimsatt, W. 1974. "Complexity and Organization." In K. Schaffner and R.S. Cohen, eds., *PSA 1972*, 69–86. Boston Studies in the Philosophy of Science, vol. 20. Dordrecht: D. Reidel.

———. 1987. "False Models as Means to Truer Theories." In M. Nitecki and A. Hoffman, eds., *Neutral Models in Biology*, 23–55. Oxford: Oxford University Press.

———. 2004. *Re-Engineering Philosophy for Limited Beings: Piecewise Approximations to Reality*. Cambridge, Mass.: Harvard University Press.

Wolff, Robert S., and Yaeger, Larry. 1993. *Visualization of Natural Phenomena*. New York: Telos (Springer-Verlag).

Wolfram, S. 1985. "Undecidability and Intractability in Theoretical Physics." *Physical Review Letters* 54, 735–738.

———. 1986. *Theory and Applications of Cellular Automata*. Singapore: World Publishing.

———. 2002. *A New Kind of Science*. Champaign, Ill.: Wolfram Media.

Wright, C. 1987. "Strict Finitism." In C. Wright, *Realism, Meaning, and Truth*, 2nd ed. Oxford: Basil Blackwell.

Wright, E.L. 1992. "Preliminary Results from the FIRAS and DIRBE Experiments on COBE." In M. Signore and C. Dupraz, eds., *The Infrared and Submillimeter Sky After COBE*. Dordrecht: Kluwer Academic.

Index

abstraction, 30, 141–147
  as property elimination, 141–143
  definition of, 143
  in models, 145
accuracy, 16–17, 21, 54
algorithm, Metropolis, 149–150
Alston, William, 10–11
amnesia, semantic, 140
argument, dilution, 12–16, 19, 22, 37
  role of identity in, 15, 37
  subject-matter-specific, 14
argument, overlap, 15–16, 17–22, 117
Armstrong, David, 45
assumptions, construction, 78, 86, 103, 128
augmentation, 4–6, 13–14, 18–19
automata, cellular, 130
axioms, 58, 96
Azzouni, Jody, 122 n.31

bridge, Millennium, 135
Bridgman, P.W., 25–28

calculation, 55
calculus, 57
calibration, 20–21, 37, 117
Carnap, Rudolf. 143
causation, property, 41–45
Chang, Hasok, 20 n.14
Church's Thesis, 123

COBE (Cosmic Background Explorer Satellite), 7, 40–41
commitment, ontological, 83
complexity theory, 70
compositional mapping, 39
computable, 50, 123–124
computers, analog, 109, 125–128
conditions
  boundary and initial, 65–66
  ceteris paribus, 87, 144
context
  of discovery, 76, 80
  of justification, 76
conversion, 4–5, 18
correction set, 78–81, 86, 103

data, quantity of, 7, 113
discovery, 25
DNA sequencing, 7
Duhem-Quine thesis, 77, 81

empiricism, 9–48, 124–125
  constructive, 11, 15
  defined, 10–11
  logical, 47
  traditional, 20, 40, 45–46, 55, 141–142
epicycles, 56, 134
equations
  differential, 60, 62, 63, 66, 68
  diffusion, 73, 80, 87, 91

equations (*continued*)
    Klein-Gordon, 70
    Laplace's, 94 n.66
    Lotka-Volterra, 63, 115
    Maxwell's, 61
    Navier-Stokes, 95–96, 117
    Schrödinger's, 60, 61, 63, 97
errors
    truncation, 118
    Type I and Type II, 48
exemplar, 101
experiment
    empirical, 115
    thought, 115–116
extrapolation, 4–5, 13, 17–18

Feller, William, 68
Feynman, Richard, 68, 140
flow, turbulent, 7, 119, 132
form, logical, 151–153
form, mathematical
    identity of, 127, 140
    vs. logical form, 151–152
Franklin, Allan, 21 n.15
Freedman, David, 114 n.14,
    118 n.20, 133

galaxies, spiral, 23–24
Galilei, Galileo, 32–33, 51
Goodman, Nelson, 143
groups, 46

Hacking, Ian, 35–36, 40
Hartmann, Stephan, 108
Harvey, William, 34
holism, epistemological.
    See Duhem-Quine thesis
hypothetico-deductivism, 81, 90

idealization, 146–147
imaging, magnetic resonance (MRI),
    24, 38
individualism, methodological, 129,
    131
instruments, off-the-shelf, 31

interpretation, detachable, 80
intuitions, 6

Kitcher, Philip, 71
knowledge, subject-specific, 14, 22,
    41, 91–95
Kuhn, Thomas, 58, 100–102

Lakatos, Imre, 55 n.14, 58, 82
Laplace, Pierre-Simon, 3, 65
law
    Newton's Second, 60–62, 65, 151
    scientific, 58, 61, 66, 88–89
Leibniz, Gottfried Wilhelm, 143
Lewis, David, 43
logic, 53, 153

machine, Turing, 52 n.11
mathematics, 53, 55, 56, 57, 89–90,
    95
    numerical, 112
'math game', 84
matter, particulate (PM), 93–94
Maxwell, Grover, 18–19
mechanics, classical, 60–62, 97
metaphor, microscope, 116–121
meteorology, 118–120
method
    in principle, 52, 55, 99, 109–110,
        114, 123, 153–156
    Monte Carlo, 118, 124, 149–150
    numerical, 51, 65, 112
model, 50, 53, 56, 58, 59, 60, 79, 101
    agent-based, 113, 129–134
    computational, 59, 102–104, 105,
        107
    false, 86, 114–115
    Ising, 63, 133, 149
    mathematical, 90

nominalism, 30, 43

observable, 9–16, 18, 23–24, 121–125
    defined, 12
observation, theory laden, 34, 39

Ohm's Law, 100–101
opacity, epistemic, 147–151
operationalism, 25–28
operations, floating point (flops), 52

paradigms, 58, 101
pattern, argument, 71
perspective, no-ownership, 77, 80, 83
photometry, 21
precision, 16–17, 21, 54
principle, property localization, 41
probability, theory of, 59, 61
process
    of adjustment, 76
    computationally irreducible, 150
    of construction, 76, 80
    Poisson, 88–89, 91, 102
programme, research, 58, 59, 82
progress, scientific, 55, 64, 154
    technology and, 120, 122
proofs, 5–6
properties, 23–28, 41–45, 91
    vs. instances, 28–30
    logical closure of, 144–145

Quine, W.V.O., 43 n.47, 83–84

realism
    property cluster, 23–25
    scientific, 155
    selective, 35–36, 82–85, 86, 122
    and underdetermination, 140–141
recognition, optical character, 16–17
representation
    dynamical, 110, 113
    form of, 98–99
    output, 103, 110–111, 113, 128
    propositional, 113–114
    in simulations, 107–109, 112
resolution, 16–17, 21, 54, 65, 118–120
Rutherford, Ernest, 41

Saturn, 33–34
science
    automated, 6–8, 112

computational, 49–55, 105–107
    limit, 154–155
    ordering of, 68–72
security, epistemic, 9, 45–48
simplification, 147
simulation, computer, 51–53, 57,
    105–135
    analog, 125–128
    core, 109–111, 113
    of data, 135
    deficiencies of, 133–135
    definition of, 110
    finite difference, 120–121
    full, 111
    as neo-Pythagoreanism, 137
    as numerical experiment, 107
    numerical mathematics and,
        112
    role of computational models in,
        109, 131
    scale, 128
    static, 110–111
skepticism, 46–47
solutions, analytic vs. numerical,
    64
sonar, side-scanning, 21–22
statistics, 57, 95
Suppes, Patrick, 97
syntax, importance of, 95–100. *See also*
    theory, syntactic account of

Tarski, Alfred, 141
telescopes, 32–34
temperature, 27–28
template
    computational, 59, 60–67, 70, 71,
        78, 95, 102
    construction of, 72–76, 81, 86, 87,
        88, 90, 140
    false, 86
    intended interpretation of, 78, 80,
        82, 103, 140
    justification of, 80, 82, 87, 103
    theoretical, 60–61, 62, 64, 84, 87, 92,
        100, 101, 103

theory, 57, 59
  semantic account of, 96, 138
  syntactic account of, 96, 138
thermometers, 27–28, 38
tomography, computerized axial
    (CAT scans), 5
tractable, 123–124
Turing, Alan, 49

underdetermination
  empirical, 137

formal, 138
  semantic, 138–139
understanding, 114, 132, 148
unobservable. *See* observable

van Fraassen, Bas, 11, 12 n.6, 15
van Helden, Albert, 32–34
visualization, computer, 112–113

Whittle, Mark, 21
Wimsatt, Bill, 85, 154 n.31

Printed in the United States
By Bookmasters